高职高专高分子材料加工技术专业规划教材编审委员会

顾　　　问　陶国良
主 任 委 员　王荣成
副主任委员　陈滨楠　陈炳和　金万祥　冉新成　王慧桂
　　　　　　　杨宗伟　周大农

委　　　员　(按姓名汉语拼音排序)

卜建新	蔡广新	陈滨楠	陈炳和	陈改荣	陈华堂
陈　健	陈庆文	丛后罗	戴伟民	邸久生	付建伟
高朝祥	郭建民	侯文顺	侯亚合	胡　芳	金万祥
孔　萍	李光荣	李建钢	李跃文	刘巨源	刘青山
刘琼琼	刘少波	刘希春	罗成杰	罗承友	麻丽华
聂恒凯	潘文群	潘玉琴	庞思勤	戚亚光	冉新成
桑　永	王国志	王红春	王慧桂	王加龙	王玫瑰
王荣成	王艳秋	王　颖	王玉溪	王祖俊	翁国文
吴清鹤	肖由炜	谢　晖	徐应林	薛叙明	严义章
杨印安	杨中文	杨宗伟	张　芳	张金兴	张晓黎
张岩梅	张裕玲	张治平	赵继永	郑家房	郑式光
周大农	周　健	周四六	朱卫华	朱　雯	朱信明
邹一明					

教育部高职高专规划教材

高分子材料专业英语

第二版

刘琼琼　主编

化学工业出版社
·北京·

内容提要

本书是《高分子材料专业英语》的第二版，是高职高专高分子材料专业的专业英语教材。全书共分为六大部分（PART A～PART F），内容包括高分子材料的基本知识（Introduction），塑料添加剂（Plastics Additives）和各种树脂（Resins），橡胶配合剂（Rubber Compounding Ingredients）和各种生胶（Raw Rubbers）以及高分子材料的成型加工（Polymer Processing）。书后附有合成橡胶的命名（The Nomenclature of Synthetic Elastomers），高聚物缩写（Acronyms of Polymers）和高分子科学术语的词汇表（Glossary of Polymer Engineering）。

本书为高职高专院校高分子材料类专业师生使用教材，也可供从事高分子材料加工及应用的专业技术人员参考。

图书在版编目 (CIP) 数据

高分子材料专业英语/刘琼琼主编.—2版.—北京：化学工业出版社，2010.7（2021.9重印）
教育部高职高专规划教材
ISBN 978-7-122-08593-1

Ⅰ.高… Ⅱ.刘… Ⅲ.高分子材料-英语-高等学校：技术学院-教材　Ⅳ.H31

中国版本图书馆 CIP 数据核字（2010）第 090386 号

责任编辑：于　卉	文字编辑：颜克俭
责任校对：蒋　宇	装帧设计：于　兵

出版发行：化学工业出版社（北京市东城区青年湖南街 13 号　邮政编码 100011）
印　　装：北京七彩京通数码快印有限公司
787mm×1092mm　1/16　印张 10¾　字数 262 千字　2021 年 9 月北京第 2 版第 7 次印刷

购书咨询：010-64518888　　　　　　　　　　　售后服务：010-64518899
网　　址：http://www.cip.com.cn
凡购买本书，如有缺损质量问题，本社销售中心负责调换。

定　　价：30.00元　　　　　　　　　　　　　　　版权所有　违者必究

前 言

  本书第一版自出版以来，得到了广大师生的肯定，并多次重印。大家普遍反映此教材区别于一般的高等学校专业英语教材，具有系统性强，篇幅较短，注释详尽的特点。

  在征求使用者意见和建议的基础上，第二版保留了第一版的特色，更换了部分难以理解的篇章，修订了部分注释内容。

  第二版改动最大的是在课文正文侧边的空白处给出了专业单词和词组的注释。改动的初衷来源于学生上课时通常将课后单词和词组注释忙于写在书中。采用此种方式可以将词汇的学习融入文章的理解之中，并且针对性较强，有利于文章的阅读。

  本书的修订工作由第一版的编写人员共同完成，主要由刘琼琼执笔。

  虽然在第二版编写时尽了最大的努力，工作量较大，但书中一定还存在一些不足之处，敬请大家将使用中发现的一些问题及时反馈，以便使教材进一步完善。

编者

2010 年 4 月

第一版前言

本书是教育部高职高专规划教材，是按照教育部对高职高专人才培养工作的指导思想，在广泛吸取了多年来专业英语教学经验的基础上编写的。

全国各高职高专学校的高分子材料专业都开设了"专业英语"课程。这门课为完成基础英语向专业英语的过渡，提高学生阅读高分子材料专业文献资料的能力发挥了重要作用，但一直以来没有相应的教材，本教材的出版将对教学起到重要的作用。

全书共分为六大部分（PART A～PART F），内容包括高分子材料的基本知识（Introduction），塑料加工中的助剂（Plastics Additives）和各种树脂（Resins），橡胶加工中的配合剂（Rubber Compounding Ingredients）和各种生胶（Raw Rubbers）以及高分子材料的成型加工（Polymer Processing）。每课均由课文（Text），单词和词组（Words and Expressions，包括音标及解释），注释（Notes），练习（Exercises）和阅读材料（Reading Material）组成。每课的课文和阅读材料均可独立成篇，取自各种不同风格的专业文献资料。具有词汇量适中、覆盖面广的特点。另外，在附录部分专列了高分子材料专业名词术语表（Glossary of Polymer Engineering）以利于学生查阅。本书主要作为高职高专教材，也可为从事高分子材料加工及应用的专业技术人员参考。

本书编写分工情况为：PART A 和 PART B 由刘晓侠编写；PART C，PART F 的 Unit 1 和 Unit 2 由高炜斌编写；PART D，PART E 和 PART F 的 Unit 3 及附录和词汇表由刘琼琼编写。全书刘琼琼主编，李永芳主审。在编写过程中得到了徐州工业职业技术学院材料工程系的各位老师给予的帮助，在此深表谢意。

限于编者水平，书中的不妥之处在所难免，希望能得到读者的批评指正以利于进一步的完善和改进。

编者
2005 年 3 月

CONTENTS

PART A Introduction
    Unit 1 Polymer 1
        [Reading Material] The Rise of Polymer Science 3
    Unit 2 Classification of Polymers 5
        [Reading Material] Polymer Architecture 8
    Unit 3 Synthesis 11
        [Reading Material] Techniques of Polymerization 15
    Unit 4 Plastics, Rubber and Fiber 17
        [Reading Material] The General Properties of Plastics 20

PART B Plastics Additives
    Unit 1 Stabilizers 22
        [Reading Material] Organic Stabilizers for Pipe 24
    Unit 2 Plasticizers 26
        [Reading Material] PVC Additives (1) 29
    Unit 3 Lubricants 31
        [Reading Material] PVC Additives (2) 33

PART C Resins
    Unit 1 Polyethylene 35
        [Reading Material] High Density Polyethylene 37
    Unit 2 Polypropylene 40
        [Reading Material] Nanometer Plastic 42
    Unit 3 Polyvinyl Chloride 45
        [Reading Material] What is PVC? (1) 47
    Unit 4 Polystyrene 49
        [Reading Material] What is PVC? (2) 51
    Unit 5 Phenolics 53
        [Reading Material] Future 15 Years in Prospect of Chemical Construction Material 55
    Unit 6 Epoxy Resins 57
        [Reading Material] Plastics and the Environment 59

PART D Rubber Compounding Ingredients
    Unit 1 General Introduction 61
        [Reading Material] Importance of Rubber 64
    Unit 2 Rubbers 66

        [Reading Material]   Other Types of Rubbers ··· 69

      Unit 3   Vulcanizing System ·· 71

        [Reading Material]   Other Types of Vulcanizing Agents ··· 74

      Unit 4   Antidegradants ··· 76

        [Reading Material]   The Ageing of Rubber Vulcanizates ··· 79

      Unit 5   Fillers ··· 81

        [Reading Material]   IRHD and Shore A Scales ·· 85

                                Silica Used in Tires ··· 85

PART E   **Raw Rubbers**

      Unit 1   Natural Rubber ··· 87

        [Reading Material]   Natural Rubber (Polyisoprene) ··· 89

      Unit 2   Butadiene Rubber ·· 90

        [Reading Material]   Difference of NR and IR ·· 93

      Unit 3   E-SBR ··· 94

        [Reading Material]   Polymerization of E-SBR ··· 97

      Unit 4   Acrylonitrile-Butadiene Rubber ·· 98

        [Reading Material]   General Types of NBR ·· 100

      Unit 5   Chloroprene Rubber ·· 102

        [Reading Material]   Rubber Compounding Rules (1) ·· 105

      Unit 6   Butyl Rubber ·· 108

        [Reading Material]   Rubber Compounding Rules (2) ·· 111

      Unit 7   Ethylene-Propylene Rubber ·· 114

        [Reading Material]   Surface Bloom of Rubber ··· 116

PART F   **Polymer Processing**

      Unit 1   Extrusion ·· 119

        [Reading Material]   Extrusion ·· 122

      Unit 2   Molding ·· 124

        [Reading Material]   Rotational, Fluidized-Bed and Slush Molding ································ 126

      Unit 3   Rubber Processing ··· 128

        [Reading Material]   Processing Rubber ··· 131

Appendix I   The Nomenclature of Synthetic Elastomers ··· 134

Appendix II   Acronyms of Polymers ··· 137

Appendix III Glossary of Polymer Engineering ··· 140

Vocabulary ·· 150

Bibliography ··· 164

PART A   Introduction

Unit 1   Polymer

A polymer is a large molecule built up from numerous smaller molecules. These large molecules may be linear, slightly branched, or highly interconnected. In the later case the structure develops into a large three-dimensional network.

The small molecules used as the basic building blocks for these large molecules are known as monomers. For example poly(vinyl chloride) is made from the monomer vinyl chloride.[1]The repeat unit in the polymer usually corresponds to the monomer from which the polymer was made. There are exceptions to this, though. Poly(vinyl alcohol) is formally considered to be made of vinyl alcohol repeat units but there is, in fact, no such monomer as vinyl alcohol. [2]To make this polymer, it is necessary first to prepare poly(vinyl acetate) from the monomer vinyl acetate, and then to hydrolyse the product to yield the polymeric alcohol.

The size of a polymer molecule may be defined either by its mass or by the number of repeat units in molecule. This later indicator of size is called the degree of polymerization, DP. The relative molar mass of the polymer is thus the product of the relative molar mass of the repeat unit and the DP.

[3]There is no clear cut boundary between polymer chemistry and the rest of chemistry. As a very rough guide, molecules of relative molar mass of at least 10000 or a DP of at least 1000 are considered to fall into the domain of polymer chemistry.

The vast majority of polymers in commercial use are organic in nature, that is, they are based on covalent compounds of carbon. The other elements involved in polymer chemistry most commonly include hydrogen, oxygen, nitrogen, chlorine, fluorine, phosphorus, sulfur and silicon, i.e. those elements which are able to form covalent bonds with carbon.

New Words

polymer	[ˈpɔlimə]	n.	聚合物
molecule	[ˈmɔlikjuːl, ˈməu-]	n.	分子
linear	[ˈliniə]	adj.	线型的
branch	[brɑːntʃ]	vt.	分支
interconnect	[ˌintə(ː)kəˈnekt]	vt.	使互相连接
network	[ˈnetwəːk]	n.	网状物
monomer	[ˈmɔnəmə]	n.	单体
vinyl	[ˈvainil, ˈvinil]	n.	乙烯基
chloride	[ˈklɔːraid]	n.	氯化物
alcohol	[ˈælkəhɔl]	n.	醇，乙醇
acetate	[ˈæsiˌteit]	n.	醋酸酯、乙酸酯，醋酸盐
hydrolyse	[ˈhaidrəlaiz]	v.	水解
product	[ˈprɔdʌkt]	n.	产品，产物，乘积
mass	[mæs]	n.	质量
indicator	[ˈindikeitə]	n.	指标
polymerization	[ˌpɔliməraiˈzeiʃən]	n.	聚合
molar	[ˈməulə]	adj.	摩尔的
boundary	[ˈbaundəri]	n.	分界线
domain	[dəuˈmein]	n.	范围，领域
organic	[ɔːˈgænik]	adj.	有机的
covalent	[kəuˈveilənt]	adj.	共价的
element	[ˈelimənt]	n.	元素，成分
carbon	[ˈkɑːbən]	n.	碳
hydrogen	[ˈhaidrəudʒən]	n.	氢
oxygen	[ˈɔksidʒən]	n.	氧
nitrogen	[ˈnaitrədʒən]	n.	氮
chlorine	[ˈklɔːriːn]	n.	氯
fluorine	[ˈflu(ː)əriːn]	n.	氟
phosphorus	[ˈfɔsfərəs]	n.	磷
sulfur	[ˈsʌlfə]	n.	硫黄
silicon	[ˈsilikən]	n.	硅

Notes

[1] The repeat unit in the polymer usually corresponds to the monomer from which the polymer was made. There are exceptions to this, though. Poly(vinyl alcohol) is formally considered to be made of vinyl alcohol repeat units but there is, in fact, no such monomer as vinyl alcohol. 聚合物的重复结构单元一般与用于制备此聚合物的单体相对应，然而，也有例外，如聚乙烯醇通常看成是由乙烯醇重复单元组成的，但实际上不存在乙烯醇这种单体。

[2] To make this polymer, it is necessary first to prepare poly(vinyl acetate) from the monomer vinyl acetate, and then to hydrolyse the product to yield the polymeric alcohol. 要制备聚乙烯醇这种聚合物，首先从单体乙酸乙烯酯制备聚乙酸乙烯酯，然后将聚乙酸乙烯酯水解得到聚乙烯醇。

[3] There is no clear cut boundary between polymer chemistry and the rest of chemistry. As a very rough guide, molecules of relative molar mass of at least 10000 or a DP of at least 1000 are considered to fall into the domain of polymer chemistry. 聚合物与其他化合物之间并无明显的界线，一般大致将相对分子质量超过10000（聚合度超过1000）的分子划为聚合物。

Exercises

1.Translate the following into Chinese

Polymers are substances containing a large number of structural units joined by the same type of linkage. The key characteristic that distinguishes polymers from other materials is their chain-like molecular structure. This structure is also responsible for the unique properties and processing behavior of polymers. Polymers in the natural world have been around since the beginning of time. Starch, cellulose, and rubber all possess polymeric properties. Man-made polymers have been studied since 1832. Today, the polymer industry has grown to be larger than the aluminum, copper and steel industries combined.

2. Give a definition for each following word

(1) polymer (2) monomer

[Reading Material]

The Rise of Polymer Science

Since most chemists and chemical engineers are now involved in some phase of polymer science or technology, some have called this the polymer age. Actually, we have always lived in a polymer age.

Polymer is derived from the Greek *poly* and *meros*, meaning many and parts, respectively. Some scientists prefer to use the word *macromolecule,* or large molecule, instead of polymer. Others maintain that naturally occurring polymers, or *biopolymers*, and synthetic polymers should be studied in different courses. However, the same principles apply to all polymers. If one discounts the end uses, the differences between all polymers, including plastics, fibers and elastomers or rubbers, are determined primarily by the intermolecular and intramolecular forces between the molecules and within the individual molecule, respectively, and by the functional group present.

Since ancient times, naturally occurring polymers have been used by mankind for various purposes. Proteins from meat and polysaccharides from grain are important sources of food. In addition to being the basis of life itself, protein, which was the first polymer, was (and is) used as a source of amino acids and energy. The ancients degraded or depolymerized the protein in tough meat by aging and cooking and denatured egg albumin by heating or adding vinegar to the eggs.

Wool and silk, both proteins, serve as clothing. Wood, the main component of which is cellulose, a polysaccharide, is used for building and fire-making. The use of asphalt as an adhesive is mentioned in the Bible. Amber, a high-molar mass resin, was worn by the Greeks as a jewel.

Early humans learned how to process, dye, and weave the natural proteinaceous fibers of wool and silk and the carbohydrate fibers of flax and cotton. Early South American civilizations used natural rubber for making elastic articles and for waterproofing of fabrics.

Nobel laureate Hermann Staudinger laid the groundwork for modern polymer science in the 1920s. The development of polymer technology since 1940s has been extremely rapid. The world production of synthetic rubber has for a long time (in 1960s) exceeded that of natural rubber and that the world production of synthetic fibers equaled to that of the natural fibers in 1970s. The volume production of plastics in the world already exceeded that of steel in the end of 20th century.

New Words and Expressions

macromolecule	[ˌmækrəuˈmɔlikjuːl]	n.	高分子，大分子
principle	[ˈprinsəpl]	n.	原理，法则，原则
elastomer	[iˈlæstəmə(r)]	n.	弹性体
intermolecular	[ˌintrəməˈlekjulə]		分子间的
intramolecular	[ˌintrəməˈlekjulə]	adj.	分子内的
functional group			官能团
polysaccharide	[pɔliˈsækəraid]	n.	多糖，聚糖，多聚糖
degrade	[diˈgreid]	v.	(使)降级
depolymerize	[diːˈpɔliməraiz]	v.	(使)解聚
denature	[diːˈneitʃə]	vt.	使变性
albumin	[ælˈbjumin]	n.	白蛋白
vinegar	[ˈvinigə]	n.	醋
cellulose	[ˈseljuləus]	n.	纤维素
asphalt	[ˈæsfælt]	n.	沥青
adhesive	[ədˈhiːsiv]	n.	胶黏剂
		adj.	黏性的
amber	[ˈæmbə]	n.	琥珀
resin	[ˈrezin]	n.	树脂
dye	[dai]	n.	染料，染色
proteinaceous	[ˌprəutiːˈneiʃəs,-tiːi-]	adj.	蛋白质的，似蛋白质的
carbohydrate	[ˈkɑːbəuˈhaidreit]	n.	碳水化合物，糖类
flax	[flæks]	n.	亚麻，麻布，亚麻织品
laureate	[ˈlɔːriit]	adj.	佩戴桂冠的
		n.	戴桂冠的人

Unit 2 Classification of Polymers

There are a number of methods of classifying polymers. [1]One is to adopt the approach of using their response to thermal treatment and to divide them into thermoplastics and thermosets. Thermoplastics are polymers which melt when heated and resolidify when cooled, while thermosets are those which do not melt when heated but, at sufficiently high temperatures, decompose irreversibly. This system has the benefit that there is a useful chemical distinction between the two groups. [2]Thermoplastics comprise essentially linear or lightly branched polymer molecules, while thermosets are substantially crosslinked materials, consisting of an extensive three-dimensional network of covalent chemical bonding.

Another classification system, first suggested by Carothers in 1929, is based on the nature of the chemical reactions employed in the polymerization. Here the two major groups are the condensation and the addition polymers. [3]Condensation polymers are those prepared from monomers where reaction is accompanied by the loss of a small molecule, usually of water. By contrast, addition polymers are those formed by the addition reaction of an unsaturated monomer, such as takes place in the polymerization of vinyl chloride.

This system was slightly modified by P. J. Flory, who placed the emphasis on the mechanisms of the polymerization reactions. He reclassified polymerizations as step reactions or chain reactions corresponding approximately to condensations or addition in Carother's scheme, but not completely. [4]A notable exception occurs with the synthesis of polyurethanes, which are formed by reaction of isocyanates with hydroxyl compounds and follow "step" kinetics, but without the elimination of a small molecule.

[5]In the first of these, the kinetics are such that there is a gradual built up of high relative molar mass material as reaction proceeds, with the highest molar mass material not appearing until the very end of the reaction. [6]On the other hand, chain reactions, which occur only at a relatively few activated sites within the reaction medium, occur with rapid build up of a few high relative molar mass molecules while the rest of the monomer remains unreacted. When formed, such macromolecules stay essentially unchanged while the rest of the monomer undergoes conversion. [7]This means that large molecules

appear very early in the polymerization reaction, which is charactered by having both high relative molar mass and monomer molecules present for most of the duration of the reaction.

New Words

classify	['klæsifai]	vt.	分类, 分等
thermal	['θə:məl]	adj.	热的, 热量的
thermoplastics	[,θə:mə'plæstiks]	n.	热塑性材料, 热塑性塑料
thermoset	['θə:məset]	n.	热固性材料, 热固性塑料
		adj.	热固的
melt	[melt]	v.	(使)熔化
solidify	[sə'lidifai]	v.	(使)凝固
decompose	[,di:kəm'pəuz]	v.	分解
comprise	[kəm'praiz]	v.	包含, 由…组成
crosslink	['krɔsliŋk]	v/n.	交联
polymerization	[,pɔlimərai'zeiʃən]	n.	聚合
condensation	[kɔnden'seiʃən]	n.	浓缩
addition	[ə'diʃən]	n.	加成
unsaturated	['ʌn'sætʃəreitid]	adj.	不饱和的
mechanism	['mekənizəm]	n.	机理, 历程
scheme	[ski:m]	n.	方案
synthesis	['sinθisis]	n.	合成
polyurethane	[,pɔli'juəriθein]	n.	聚氨酯
isocyanate	[,aisəu'saiəneit]	n.	异氰酸酯
hydroxyl	[hai'drɔksil]	n.	羟基
kinetics	[kai'netiks]	n.	动力学
elimination	[i,limi'neiʃən]	n.	除去, 消除
activated	['æktiveitid]	adj.	有活性的
medium	['mi:djəm]	n.	介质
unreacted	['ʌnri'æktid]	adj.	未反应的
undergo	[,ʌndə'gəu]	vt.	经历
conversion	[kən'və:ʃən]	n.	转化
duration	[djuə'reiʃən]	n.	持续时间, 为期

Notes

[1] One is to adopt the approach of using their response to thermal treatment and to divide them into thermoplastics and thermosets. Thermoplastics are polymers which melt when heated and resolidify when cooled, while thermosets are those which do not melt when heated but, at sufficiently high temperatures, decompose irreversibly. 一种是根据热行为将聚合物分成热塑性和热固性。热塑性聚合物加热熔化, 冷却固化, 而热固性聚合物加热不会熔化, 在温度

特别高时则会发生不可逆分解。

[2] Thermoplastics comprise essentially linear or lightly branched polymer molecules, while thermosets are substantially crosslinked materials, consisting of an extensive three-dimensional network of covalent chemical bonding. 热塑性聚合物基本上是由线性的或支化的高分子组成；固性聚合物则是完全交联的，它是通过共价键连接而成的三维网状结构材料。

[3] Condensation polymers are those prepared from monomers where reaction is accompanied by the loss of a small molecule, usually of water. By contrast, addition polymers are those formed by the addition reaction of an unsaturated monomer, such as takes place in the polymerization of vinyl chloride. 缩聚物是由单体通过缩合反应形成，反应过程中有小分子生成，通常是水。与缩聚物不同，加聚物是由不饱和单体通过加聚反应形成，例如氯乙烯单体的聚合。

[4] A notable exception occurs with the synthesis of polyurethanes, which are formed by reaction of isocyanates with hydroxyl compounds and follow "step" kinetics, but without the elimination of a small molecule. （上述两种分类方法不一致）在聚氨酯的合成表现明显，聚氨酯是由异氰酸酯与羟基化合物反应，遵循逐步反应机理，但反应过程中却没有小分子生成。

[5] In the first of these, the kinetics are such that there is a gradual built up of high relative molar mass material as reaction proceeds, with the highest molar mass material not appearing until the very end of the reaction. 第一类聚合（逐步聚合）的机理是随着反应的进行，分子量逐步增加，直到反应结束才有高分子生成。

[6] On the other hand, chain reactions, which occur only at a relatively few activated sites within the reaction medium, occur with rapid build up of a few high relative molar mass molecules while the rest of the monomer remains unreacted. 另一类连锁聚合反应只是在少量活性中心上进行，反应体系中迅速生成高分子，同时体系中还存有未反应的单体。

[7] This means that large molecules appear very early in the polymerization reaction, which is charactered by having both high relative molar mass and monomer molecules present for most of the duration of the reaction. 这意味着高分子在聚合反应一开始就会形成，表现为几乎在整个聚合反应期间高分子和单体分子同时存在。

Exercises

Read the following short article and comprehend the classificaiton of polymers:

1) **Natural – Synthetic**
Based on the origin of the material, whether natural or synthesized.

2) **Organic – Inorganic**
Organic polymers have carbon backbone. PE, PP, PS etc.
Inorganic polymers do not contain carbon backbone. Glass, Silicone polymers.

3) **Thermoplastic – Thermosetting**
Thermoplastics soften on heating.
Thermosets do not soften or melt on heating. Cross-linked chains.

4) **Plastics, elastomers, fibers, resins**
Classified as per the use of polymeric material.

5) **Addition polymers – Condensation polymers**

Based on manufacturing process.

The classification based on manufacturing process is important for the course, "Industrial Polymerization".

Addition Polymerization: Also called as *Chain Polymerization, Chain growth Polymerization, Chain Reaction Polymerization.*

Condensation Polymerization: Also called as *Step Polymerization, Step Growth Polymerization, Step Reaction Polymerization.*

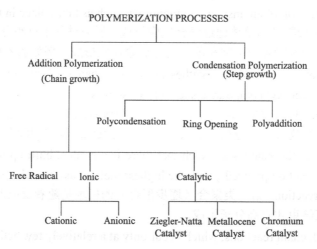

Examples:

Addition polymers: Polyethylene, Polypropylene, PVC, Polystyrene.

Condensation polymers: Polyesters, Nylon 66, Polycarbonates, Polyurethanes, Epoxy resins.

[Reading Material]

Polymer Architecture

An important microstructural feature determining polymer properties is the polymer architecture. The simplest polymer architecture is a linear chain: a single backbone with no branches. A related unbranching architecture is a ring polymer. A branched polymer molecule is composed of a main chain with one or more substituent side chains or branches. Special types of branched polymers include star polymers, comb polymers, brush polymers, dendronized polymers, ladders, and dendrimers.

Branching of polymer chains affects the ability of chains to slide past one another by altering intermolecular forces, in turn affecting bulk physical polymer properties. Long chain branches may increase polymer strength, toughness, and the glass transition temperature due to an increase in the number of entanglements per chain. The effect of such long-chain branches on the size of the polymer in solution is characterized by the branching index. Random length and atactic short chains, on the other hand, may reduce polymer strength due to disruption of organization and may likewise reduce the crystallinity of the polymer.

A good example of this effect is related to the range of physical attributes of polyethylene. High-density polyethylene (HDPE) has a very low degree of branching, is quite stiff, and is used in applications such as milk jugs. Low-density polyethylene (LDPE), on the other hand, has significant numbers of both long and short branches, is quite flexible, and is used in applications such as plastic films.

Dendrimers are a special case of polymer where every monomer unit is branched. This tends to reduce intermolecular chain entanglement and crystallization. Alternatively, dendritic polymers are not perfectly branched but share similar properties to dendrimers due to their high degree of branching.

The architecture of the polymer is often physically determined by the functionality of the monomers from which it is formed. This property of a monomer is defined as the number of reaction sites at which may form chemical covalent bonds. The basic functionality required for forming even a linear chain is two bonding sites. Higher functionality yields branched or even crosslinked or networked polymer chains.

An effect related to branching is chemical crosslinking - the formation of covalent bonds between chains. Crosslinking tends to increase T_g and increase strength and toughness. Among other applications, this process is used to strengthen rubbers in a process known as vulcanization, which is based on crosslinking by sulfur. Car tires, for example, are highly crosslinked in order to reduce the leaking of air out of the tire and to toughen their durability. Eraser rubber, on the other hand, is not crosslinked to allow flaking of the rubber and prevent damage to the paper.

A crosslink suggests a branch point from which four or more distinct chains emanate. A polymer molecule with a high degree of crosslinking is referred to as a polymer network. Sufficiently high crosslink concentrations may lead to the formation of an infinite network, also known as a gel, in which networks of chains are of unlimited extent-essentially all chains have linked into one molecule.

Words and Expressions

architecture	[ˈɑːkitektʃə]	n.	构造
ring	[riŋ]	n.	链状
side chain			侧链，支链
star polymer			星形高聚物
comb polymer			梳型高聚物
dendronized polymer			树形高分子
ladder	[ˈlædə]	n.	梯形高分子
dendrimer		n.	超支化高分子
strength	[streŋθ]	n.	强度
toughness	[ˈtʌfnis]	n.	韧性
entanglement	[inˈtæŋglmənt]	n.	缠结
crystallinity	[ˌkristəˈliniti]	n.	结晶度
high-density polyethylene (HDPE)			高密度聚乙烯
degree of branching			支化度
milk jug		n.	奶壶

low-density polyethylene (LDPE)　　　　　　　　　　低密度聚乙烯
film   [film]   n.   薄膜
functionality   [ˌfʌŋkəʃəˈnæliti]   n.   官能度
vulcanization   [ˌvʌlkənaiˈzeiʃən; -niˈz-] n.   硫化
emanate   [ˈeməneit]   vt.   发出，发源
degree of crosslinking   交联度
crosslink concentration   交联密度
gel   [dʒel]   n.   凝胶

Unit 3　Synthesis

Step-growth Polymerization

It is conventional to divide the synthesis of polymers into two main categories. [1]One is step-growth polymerization which is often also called *condensation polymerization* since it is almost exclusively concerned with condensation reactions taking place between multifunctional monomer molecules. The other category is *addition polymerization* where the monomer molecules add on to a growing chain one at a time and no small molecules are eliminated during the reaction. [2]Step-growth polymerization is characterized by the gradual formation of polymer chains through successive reactions coupling monomers to each other to form dimers which can also react with other dimers or unreacted monomer molecules.

Since step-growth polymerization is mainly concerned with condensation reactions, it will be useful first of all to look at what is meant by a condensation reaction. In simple terms it is a reaction between an organic base (such as an alcohol or amine) with an organic acid (such as a carboxylic acid or acid chloride) in which a small molecule like water is eliminated (condensed out).

A typical condensation reaction is the reaction between acetic acid and ethyl alcohol:

$$CH_3COOH + C_2H_5OH \longrightarrow CH_3COOC_2H_5 + H_2O$$

Two important points to note are that a small molecule, water, is produced by the reaction. Also the product ethyl acetate is known as an *ester*. There are many examples of condensation reactions in organic chemistry and other products as well as esters are produced. For example, amides can be produced by reaction between organic acids and amines.

Free Radical Addition Polymerization

Addition polymerization is the second main type of polymerization reaction. It differs from step-growth polymerization in several important ways. [3]It takes place in three distinct steps *initiation, propagation and termination* and the principal mechanism of polymer formation is by addition of monomer molecules to a growing chain. Monomers for addition polymerization normally contain double bonds and are of the general formula $CH_2=CR_1R_2$. The double bond is susceptible to attack by either underlined free radical or ionic underlined initiators to form a species known as an *active center*. This active center propagates a growing chain by the addition of monomer molecules and the active center is eventually neutralized by a termination reaction. [4]Since the reaction only occurs at the reactive end of the growing chain, long molecules are present at an early stage in the reaction along with unreacted monomer molecules which are around until very near the end of the reaction. It is possible to generalize about many feature of addition polymerization, but there are sufficient differences between free radical and ionic initiated reactions for the two types to be treated separately.

Ionic Polymerization

Ionic polymerizations are classified according to whether the polymeric ions are positively or negatively charged. If the ions are positively charged the polymerization is termed *cationic* and if they negatively charged it is known as *anionic* polymerization. It is normally possible to generalize about the mechanisms and kinetics of free-radical addition polymerization, but in the case of ionic initiated polymerization such generalizations cannot normally be made. Unlike free-radical systems the detailed mechanisms depend upon the type of initiator, monomer and solvent employed. When the chain carriers are ionic the addition reactions are often rapid and difficult to reproduce. Also high rates and degrees of polymerization can be obtained at low temperatures Propagation rarely takes place through free ions. Normally there are counter-ions closely associated with the polymer ions during chain propagation. The nature of the solvent influences the strength of the association and the distance between the ions in the ion-pair. The presence of the counter-ion, in turn, can control the stereochemistry and the rate of monomer addition.

Ionic polymerization systems have been developed because some monomers which contain double bonds cannot be polymerized using free-radical initiators. Also ionic polymerization generally takes place at low temperatures and can offer better control of stereoregularity and relative molecular mass distribution.

Copolymerization

It is common practice in materials science to improve the properties of one material by mixing it in some way with another material. Metal alloys have mechanical properties which are vastly superior to those of pure metals. In polymer science this is rather more difficult to do as few polymers can be mixed or blended satisfactorily and the properties of the blend are often inferior to those of the pure polymers. An alternative approach which can be adopted with polymers is to synthesize chains which have more than one type of monomer unit. [5]<u>Copolymers</u> which are made by this method can have qualities which may be better than those of the parent <u>homopolymers</u>, but the properties of the copolymer are usually characterized by the detailed arrangements of the monomers units on the polymer chain.

共聚物
均聚物

New Words

synthesis	['sinθisis]	n.	合成
condensation	['kɔnden'seiʃən]	n.	缩合
addition	[ə'diʃən]	n.	加成
successive	[sək'sesiv]	adj.	连续的
dimer	['daimə]	n.	二聚物
base	[beis]	n.	碱
alcohol	['ælkəhɔl]	n.	醇
amine	['æmi:n]	n.	胺
carboxylic	[,kɑ:bɔk'silik]	adj.	羧基的
chloride	['klɔ:raid]	n.	氯化物
acetic	[ə'si:tik]	adj.	醋酸的, 乙酸的
ester	['estə]	n.	酯
amide	['æmaid]	n.	酰胺
initiation	[i,niʃi'eiʃən]	n.	引发
propagation	[,prɔpə'geiʃne]	n.	增长
termination	[,tə:mi'neiʃən]	n.	终止
mechanism	['mekənizəm]	n.	机理
formula	['fɔ:mjulə]	n.	化学式, 公式
susceptible	[sə'septəbl]	adj.	易受影响的
ionic	[ai'ɔnik]	adj.	离子的
initiator	[i'niʃieitə]	n.	引发剂
neutralize	['nju:trəlaiz; (US)nu:-]	vt.	压制
cation	['kætaiən]	n.	阳离子
anionic	['ænaiənik]	adj.	阴离子的
kinetics	[kai'netiks]	n.	动力学
solvent	['sɔlvənt]	n.	溶剂
stereochemistry	[,stiəriə'kemistri]	n.	立体化学
stereoregularity	[,stiəriə,regju'læriti]		立构规整性
blend	[blend]	vt./ n.	混合, 共混
copolymerization	[kəu,pɔlimərai'zeiʃən]	n.	共聚合
copolymer	[kəu'pɔlimə]	n.	共聚物
homopolymer	[,həumə'pɔlimə]	n.	均聚物

Notes

[1] It is conventional to divide the synthesis of polymers into two main categories. One is step-growth polymerization which is often also called *condensation polymerization* since it is almost exclusively concerned with condensation reactions taking place between multifunctional monomer molecules. 聚合反应通常分成两大类，一类是逐步聚合，逐步聚合也称为缩合聚合，因为这种反应就是含多个官能团的单体之间进行的缩合反应。

[2] Step-growth polymerization is characterized by the gradual formation of polymer chains through successive reactions coupling monomers to each other to form dimers which can also react with other dimers or unreacted monomer molecules. 逐步聚合的特征表现为高分子长链是逐渐形成的，两两单体先形成二聚体，这个二聚体又和其他二聚体或未反应的单体连续不断反应，最后形成高分子链。

[3] It takes place in three distinct steps *initiation, propagation and termination* and the principal mechanism of polymer formation is by addition of monomer molecules to a growing chain. 加成聚合反应分成明显的三步：引发、增长和终止。形成高分子的主要机理是单体不断加成到增长的活性长链上。

[4] Since the reaction only occurs at the reactive end of the growing chain, long molecules are present at an early stage in the reaction along with unreacted monomer molecules which are around until very near the end of the reaction. 尽管反应仅发生在增长活性长链的末端，在反应早期仍然有高分子生成，同时一直到反应结束体系中都存在未反应的单体。

[5] Copolymers which are made by this method can have qualities which may be better than those of the parent homopolymers, but the properties of the copolymer are usually characterized by the detailed arrangements of the monomers units on the polymer chain. 通过共聚所得到的共聚物的性能比原来的均聚物的性能好，而共聚物的性能通常是由高分子链中的单体单元的具体排列方式表现出来的。

Exercises

1.Translate the following into Chinese

Branching

During the propagation of polymer chains, branching can occur. In radical polymerization, this is when a chain curls back and bonds to an earlier part of the chain. When this curl breaks, it leaves small chains sprouting from the main carbon backbone. Branched carbon chains cannot line up as close to each other as unbranched chains can. This causes less contact between atoms of different chains, and fewer opportunities for induced or permanent dipoles to occur. A low density results from the chains being further apart. Lower melting points and tensile strengths are evident, because the intermolecular bonds are weaker and require less energy to break.

2. Put the following words or expressions into Chinese

单体	引发剂	分子量	自由基聚合	逐步聚合
聚合物	引发	分子量分布	阴离子聚合	连锁聚合
均聚物	增长	立构规整性	阳离子聚合	缩合聚合

共聚物　　　　终止　　　　　动力学　　　　　共聚合　　　　　加成聚合

3. Please write out at least 6~8 kinds of polymers both in English and in Chinese

[Reading Material]

Techniques of Polymerization

A number of chain-growth polymers are commercially produced on a large scale, and the technology of polymerization deserves some comment. Many important monomers polymerize with the evolution of large amounts of heat. This can result in temperature increases, increased kinetic constants, and accelerated reaction rates - in short, to runaway reactions - unless the heat is dissipated. Many important monomers are toxic, carcinogenic, or both and hence must be processed carefully. These facts, plus the diversity of monomers employed and the assortment of end uses for polymer products, make the choice of a polymerization technique a highly specific matter. In this section we shall discuss some of the possibilities.

Several polymerization techniques are in widespread usage. Our discussion is biased in favor of methods that reveal additional aspects of addition polymerization and not on the relative importance of methods in industrial practice. We shall discuss four polymerization techniques: bulk, solution, suspension, and emulsion polymerization.

A bulk polymerization may be conducted with as few as two components: monomer and initiator. Production polymerization reactions are carried out to high conversions which produces several consequences.

Solution polymerization is the name given to the technique of polymer formation in the presence of a solvent. The polymer may not have the same solubility in the solvent as the monomer, so the system may become heterogeneous with the formation of polymer. Although there is an increase in viscosity as polymer concentration in the system increases, the effect is mitigated by the presence of solvent, and autoacceleration through the Trommsdorff mechanism is less troublesome than in bulk polymerizations. This permits easier stirring and better beat exchange than in the bulk. The solvent is ordinarily chosen to show no chain transfer so that high molecular weight polymer is obtained. Solvent recovery must be considered as part of the overall process and this, of course, adds to the cost of this method.

A technique called suspension polymerization is sometimes employed with water-insoluble monomers. In this procedure the monomer is suspended by agitation as small drops in an aqueous medium. The diameters of the drops are in the range of micrometers to millimeters in this technique. Coalescence of the monomer drops is prevented by stirring and by addition of water-soluble polymers like gelatin or by suspending clay particles like kaolin in the mixture.

Oil-soluble initiators are used so that each monomer drop behaves as a miniature bulk polymerization system. The advantage of suspension polymer is the ease of heat removal, and a disadvantage is the need to separate the polymer from the suspending medium and wash it free of additives. Suspension polymerization is also called pearl polymerization because of the appearance of the polymer produced.

The fourth and most interesting of the polymerization techniques we shall consider is called emulsion polymerization. It is important to distinguish between suspension and emulsion polymerization, since there is a superficial resemblance between the two and their terminology has potential for confusion: A suspension of oil drops in water is called an emulsion. Water-insoluble monomers are used in he emulsion process also, and the polymerization is carried out in the presence of water.

Emulsion polymerization also has the advantages of good heat transfer and low viscosity, which follow from the presence of the aqueous phase. The resulting aqueous dispersion of polymer is called latex. The polymer can be subsequently separated from the aqueous portion of the latex or the latter can be used directly in eventual applications. For example, in coatings applications such as paints, paper coatings, floor polishes-soft polymer particles coalesce into a continuous film with the evaporation of water after the latex has been applied to the substrate.

Words and Expressions

constant	['kɔnstənt]	n.	常数, 恒量
		adj.	不变的
dissipate	['disipeit]	v.	驱散, (使)消散
toxic	['tɔksik]	adj.	有毒的
carcinogenic	[kɑ:sinə'dʒənik]	adj.	致癌物(质)的
assortment	[ə'sɔ:tmənt]	n.	分类
bulk	[bʌlk]	n.	大批, 本体
solution	[sə'lju:ʃən]	n.	溶液, 溶解
suspension	[səs'penʃən]	n.	悬浮, 悬浮液
emulsion	[i'mʌlʃən]	n.	乳液
solubility	[ˌsɔlju'biliti]	n.	溶解性, 溶度
heterogeneous	[ˌhetərəu'dʒi:niəs]	adj.	非均相的
viscosity	[vis'kɔsiti]	n.	黏度, 黏性
concentration	[ˌkɔnsen'treiʃən]	n.	浓度, 浓缩
autoacceleration		n.	自加速
stir	[stə:]	vt.	搅动, 搅拌
agitation	[ædʒi'teiʃən]	n.	搅拌
aqueous	['eikwiəs]	adj.	水的, 水溶液的
micrometer	[mai'krɔmitə(r)]	n.	微米
coalescence	[ˌkəuə'lesns]	n.	合并, 凝结
gelatin	['dʒelətin, 'dʒelə'ti:n]	n.	明胶, 凝胶
kaolin	['keiəlin]	n.	高岭土, 瓷土
miniature	['minjətʃə]	n.	缩影
		adj.	微型的, 缩小的
superficial	[sju:pə'fiʃəl]	adj.	表面的, 肤浅的
resemblance	[ri'zembləns]	n.	类同之处

Unit 4　Plastics, Rubber and Fiber

There are five major applications for polymers: plastics, rubbers, synthetic fibers, coatings and adhesives. The properties of the polymers are undoubtedly the most important in determining the ultimate application. [1]Polymers are rarely used in chemically pure form, so in a discussion of the technology of polymers, it is necessary to mention, in addition to the properties of the polymers required for a given application, the nature and reason for use of the many other materials often associated with the polymers.

Plastics are normally thought of as being polymer compounds possessing a degree of structural rigidity – in terms of the usual stress-strain test, a modulus on the order of 10^9 dynes/cm^2 (1dyne/cm^2=0.1Pa) or greater. The molecular requirements for a polymer to be used in a plastic compound are (a) if linear or branched, the polymer must be below its glass-transition temperature (if amorphous) and/or below its crystalline melting point (if crystallizable) are use temperature; otherwise (b) it must be crosslinked sufficiently to restrict molecular response essentially to straining of bond angles and lengths (e.g. ebonite or "hard" rubber). (c) in addition to the polymer itself, which is seldom used alone, plastics usually contain at least small amounts of one or more of the following additives: reinforcing fillers, pigments, dyes, plasticizers, lubricants and processing aids.

[2]A rubber is generally defined as a material that can be stretched to at least twice its original length and that will retract rapidly and forcibly to substantially its original dimensions on release of the force. An elastomer is a rubberlike material from the standpoint of modulus but one that has limited extensibility and incomplete retraction. The most common example is highly plasticized polyvinyl chloride.

From a molecular standpoint, a rubber must be (a) a high polymer, since rubber elasticity is due mainly to the uncoiling and coiling of long chains. In order for the molecules to be able to coil and uncoil freely, (b) the polymer must be amorphous in its unstretched state, as crystallization would hinder the molecular motion necessary for rubber elasticity. Until fairly recently, an additional requirement was (c) that the polymer be crosslinked. If this were not so, the chains would slip past one another under stress (viscous flow) and recovery would be incomplete.

While many of the polymers used for synthetic fibers are identical to those in plastics, the two industries grew up separately, with completely different <u>terminologies</u>, <u>testing procedures</u>, and so on. 术语/测试方法
[3]Many of the requirements for fabrics are stated in nonquantitative terms such as "<u>hand</u>" and "<u>drape</u>" which are difficult to relate to normal 手感/悬垂性
<u>physical property</u> measurements, but which can be critical from the 物理性能
standpoint of consumer acceptance.

A fiber is often defined as an object with a <u>length-to-diameter ratio</u> 长径比
of at least 100. Synthetic fibers are spun in the form of continuous filaments but may be chopped up to much shorter <u>staple</u>, which is then 短纤维
<u>twisted</u> into thread before <u>weaving</u>. <u>Natural fibers</u>, with the exception of 捻/织/天然纤维
silk, are initially in staple form. The thickness of a fiber is most commonly expressed in terms of <u>denier</u>, which is the weight in grams of 旦尼尔（旦数）
a 9000m length of fiber. Stresses and tensile strengths are reported in terms of <u>tenacity</u>, with <u>units of grams per denier</u>. 强力/单位是(克/旦尼尔)

New Words

application	[ˌæpliˈkeiʃən]	n.	应用，申请，运用
rigidity	[riˈdʒiditi]	n.	坚硬，刚性，刚度
stress	[stres]	n.	应力
strain	[strein]	n.	应变
modulus	[ˈmɔdjuləs]	n.	模量
amorphous	[əˈmɔːfəs]	adj.	无定形的，非晶的
crystalline	[ˈkristəlain]	adj.	结晶的
ebonite	[ˈebənait]	n.	硬橡胶
additive	[ˈæditiv]	n.	添加剂
reinforce	[ˌriːinˈfɔːs]	vt.	补强，加固
filler	[ˈfilə]	n.	填料
pigment	[ˈpigmənt]	n.	颜料
plasticizer	[ˈplæstisaizə]	n.	增塑剂
lubricant	[ˈluːbrikənt]	n.	滑润剂
forcibly	[ˈfɔːsəbli]	adv.	强制地，有力地
standpoint	[ˈstændpoint]	n.	立场，观点
extensibility	[iksˌtensəˈbiliti]	n.	延长性，伸长
elasticity	[ilæsˈtisiti]	n.	弹性
quantitative	[ˈkwɔntitətiv]	adj.	定量的
hand	[hænd]	n.	手感
drape	[dreip]	n.	悬垂性

ratio	['reiʃiəu]	n.	比，比率
spin（span spun）	[spin]	v.	纺，纺纱
filament	['filəmənt]	n.	细丝
chop	[tʃɔp]	vt.	切，砍
staple	['steipl]	n.	短纤维
twist	[twist]	vt.	捻，编织
denier	[di'naiə]	n.	旦尼尔
tenacity	[ti'næsiti]	n.	强力

Notes

[1] Polymers are rarely used in chemically pure form, so in a discussion of the technology of polymers, it is necessary to mention, in addition to the properties of the polymers required for a given application, the nature and reason for use of the many other materials often associated with the polymers. 聚合物很少单独使用，因而在讨论聚合物的技术性问题时，除了要考虑应用时的性能要求外还必须说明加入到聚合物的其他材料的性能及原因。

[2] A rubber is generally defined as a material that can be stretched to at least twice its original length and that will retract rapidly and forcibly to substantially its original dimensions on release of the force. 橡胶通常定义为一种能在拉伸后伸长到至少原长的两倍而除去外力后又能迅速有力并充分恢复到原长的材料。

[3] Many of the requirements for fabrics are stated in nonquantitative terms such as "hand" and "drape" which are difficult to relate to normal physical property measurements, but which can be critical from the standpoint of consumer acceptance. 对纤维的诸多要求都是采用非定量的术语例如"手感"、"悬垂感"，这些很难与常规的物性测试联系起来，但从消费者接受这一角度来说是很重要的。

Exercises

1. Translation the following passage into Chinese

Most conveniently, polymers are generally subdivided in three categories, viz., plastics, rubbers and fibers. In terms of initial elastic modulus, rubbers ranging generally between 10^6 to 10^7 dynes/cm^2, represent the lower end of the scale, while fibers with high initial moduli of 10^{10} to 10^{11} dynes/cm^2 are situated on the upper end of the scale; plastics, having generally an initial elastic modulus of 10^8 to 10^9 dynes/cm^2, lie in-between. As is found in all phases of polymer chemistry, there are many exceptions to this categorization.

An elastomer (or rubber) results form a polymer having relatively weak interchain forces and high molecular weights. When the molecular chains are "straightened out" or stretched by a process of extension, they do not have sufficient attraction for each other to maintain the oriented state and will retract once the force is released. This is the basis of elastic behavior.

2. Translate the second and fourth paragraph of the text

3. Give a definition for each following word

　(1) plastics　(2) rubber

[Reading Material]

The General Properties of Plastics

Plastics have many desirable properties, as shows below.

Weathering resistance: Many plastics resist weather change due to low coefficients of expansion and contraction.

Chemical resistance: Owing to their noncorrosive properties, many plastics can resist most acids, alkalis, salts, solvents, chemicals, oils and so on.

Dielectric properties: Plastics have good insulating properties with dielectric constant varying from below 2.0 to 10 above.

Finish: Virtually any finish can be obtained in a plastic part. The plastics material will reproduce a permanent finish from the finish of the mold steel.

Strength: plastics have good strength characteristics. Some reinforced plastics may be as strong as steel.

Specific gravity: plastics have specific gravities ranging 0.9 to 2.3, about one half the weight aluminum and one sixth that of steel. Some are as light as feather.

Abrasion resistance: There are plastics that will resist wear better than most metals in many applications.

Color: Plastics can be molded in many colors. The color is permanent.

Configuration: Plastics can be molded, extruded, thermal formed, or machnined into many intricate shapes, with easy methods, low costs and short cycle comparatively.

It is certainly the case that no plastic is as strong as steel, and the best elastic modulus of plastic is an order of magnitude below that of the weakest metal. No plastic is as transparent as glass. And none have the coefficients of thermal expansion possessed by metals. And because of the organic nature of the plastic materials, none have the useful temperature range of ceramics or metals. No plastic can match the abrasion resistance of porcelain and none can match the fire retardance of asbestos. Only a few plastics can meet the electrical resistance of rubber or mica, and certainly none can meet the conductivity of copper. Nevertheless, there are many applications where these extreme conditions are not needed and plastics can substitute readily. For example, in surgical implant applications, the high-temperature performance of stainless steel is not needed, whereas its corrosion resistance and chemical inertness are. Replacement of the very rigid metal with a more malleable plastic having equivalent corrosion resistance and chemical inertness can be accomplished readily. Cast iron has replaced with plastic in pump housings not because of operating temperature constraints but because plastics are not as susceptible to fatigue or corrosion. Synthetic fabrics are known for their wearability and colorfastness, even though cotton is considerably more absorbant. Therefore, proper selection of the plastics materials properties to meet the specific need is required, in the same way a suitable metal is selected to meet specific conductive or thermal needs.

Words and Expressions

weathering resistance			耐候性
coefficient	[kəuiˈfiʃənt]	n.	系数
alkali	[ˈælkəlai]	n.	碱

		adj.	碱性的
chemical	['kemikəl]	adj.	化学的
		n.	化学药品
dielectric	[ˌdaii'lektrik]	n.	电介质，绝缘体
dielectric constant			介电常数
finish	['finiʃ]	n.	光洁度
permanent	['pə:mənent]	adj.	永久的，持久的
specific gravity			相对密度，比重（旧称）
aluminum	[ə'lju:minəm]	n.	铝
abrasion	[ə'breiʒən]	n.	磨损
configuration	[kənˌfigju'reiʃən]	n.	构造，结构，构型
mold	[məuld]	n.	模型
		vt.	模塑，模压
extrude	[eks'tru:d]	v.	挤出
thermal form			热成型
intricate	['intrikit]	adj.	复杂的，错综的
magnitude	['mægnitju:d]	n.	大小，数量，量级
transparent	[træns'pɛərənt]	adj.	透明的
ceramic	[si'ræmik]	n.	陶瓷
		adj.	陶器的
porcelain	['pɔ:slin, -lein]	n.	瓷器，瓷
fire retardance			阻燃性
asbestos	[æz'bestɔs]	n.	石棉
mica	['maikə]	n.	云母
conductivity	[ˌkɔndʌk'tiviti]	n.	电导率，传导率，传导性
surgical	['sə:dʒikəl]	adj.	外科的
		n.	外科手术
implant	[im'plɑ:nt]	v.	植入
inertness		n.	不活泼，惰性
malleable	['mæliəbl]	adj.	有延展性的
cast iron			铸铁
constraint	[kən'streint]	n.	约束，强制
fatigue	[fə'ti:g]	n.	疲乏，疲劳
wearability	[ˌweərə`biləti]	n.	穿着性能，耐磨损性
colorfast	[´kʌləfɑ:st]	adj.	不退色的
absorbent	[əb`sɔ:bənt]	adj.	能吸收的
		n.	吸收剂

PART B   Plastics Additives

Unit 1   Stabilizers

Stabilizers provide protection against degradation caused by heat, oxidation, and solar radiation. Thus, when used in plastic compositions they may be classified as heat (or thermal) stabilizers, antioxidants and UV light stabilizers.

[1] It is the role of heat stabilizers to prevent the polymer from degrading during the short period of high temperature (150℃ to 300℃) processing and to protect the finished plastic article against slow aging over longer periods at service temperatures.

Antioxidants inhibit or retard oxidative degradation at normal or elevated temperatures during processing, storage or service. Most polymers undergo some oxidative degradation, but hydrocarbon polymers are specially susceptible. Antioxidants. therefore, are generally added in small quantities.

Most plastics exhibit varying degrees of degradation upon prolonged outdoor exposure. Polypropylene, poly(vinyl chloride), polyethylene, polyesters, crystalline and high impact polystyrenes, and ABS are particularly sensitive. Other plastics, particularly poly(methyl methacrylate) and the fluorocarbons, are much more resistant. To arrest or retard polymer degradation caused by the ultraviolet portion of solar radiation, plastic formulations contain UV absorbers. These are compounds such as substituted benzophenones, benzotriazoles and acrylonitriles that selectively absorb harmful radiation and convert it to heat energy.

Pigments such as titanium dioxide and zinc oxide are also used to protect plastics against the harmful effect of ultraviolet radiation.

[2] They function by absorbing some UV radiation, but their ability to reflect radiation (heat as well as light) accounts for much of their effectiveness. In applications where colour is not a requirement, carbon black, which absorbs UV light, is widely used as a very effective

stabilizer (e.g., in black polyethylene).

New Words

stabilizer	[ˈsteibilaizə]	n.	稳定剂
degradation	[ˌdegrəˈdeiʃən]	n.	降解
oxidation	[ɔksiˈdeiʃən]	n.	氧化
radiation	[ˌreidiˈeiʃən]	n.	辐射，放射
composition	[kɔmpəˈziʃən]	n.	成分
thermal	[ˈθəːməl]	adj.	热的，热量的
antioxidant	[ˈæntiˈɔksidənt]	n.	抗氧剂
aging	[ˈeidʒiŋ]	n.	老化
inhibit	[inˈhibit]	v.	抑制
retard	[riˈtɑːd]	vt.	延迟，阻止
hydrocarbon	[ˈhaidrəuˈkɑːbən]	n.	烃，碳氢化合物
polypropylene	[ˌpɔliˈprəupiliːn]	n.	聚丙烯
polyester	[ˈpɔliestə]	n.	聚酯
crystalline	[ˈkristəlain]	adj.	结晶的
impact	[ˈimpækt]	n.	冲击
polystyrene	[ˌpɔliˈstaiəriːn]	n.	聚苯乙烯
methyl	[ˈmeθil, ˈmiːθail]	n.	甲基
methacrylate	[meˈθækrəleit]	n.	甲基丙烯酸酯（盐）
fluorocarbon	[ˌflu(ː)ərəˈkɑːbən]	n.	碳氟化合物
substituted		adj.	取代的
benzophenone	[benzəufiˈnəun, -ˈfiːnəun]	n.	苯甲酮
benzotriazole	[ˌbenzəuˌtraiˈæzəul]		苯并三唑
acrylonitrile	[ˌækrələuˈnaitril]	n.	丙烯腈
titanium	[taiˈteinjəm, ti-]	n.	钛
dioxide	[daiˈɔksaid]	n.	二氧化物
zinc	[ziŋk]	n.	锌

Notes

[1] It is the role of heat stabilizers to prevent the polymer from degrading during the short period of high temperature (150℃ to 300℃) processing and to protect the finished plastic article against slow aging over longer periods at service temperatures. 热稳定剂的作用是防止聚合物在加工时高温(150~300℃)短时间内的降解和防止塑料制品在使用温度下的长时间的缓慢老化。

[2] They function by absorbing some UV radiation, but their ability to reflect radiation (heat as well as light) accounts for much of their effectiveness. 它们（颜料）可以通过吸收部分紫外线的辐射从而起到抗紫外线的作用，但它们主要还是靠对辐射（还有热和光）的反射来达到抗紫外的效果。

Exercises

1. Translate the following passage into Chinese

Heat and light stabilizers top the list of compounds most often used to upgrade recycled resins. One reason is that stabilizers tend to be decomposed or dissolved during cleaning, drying, compounding, and other operations that accompany resin reprocessing. In addition, earlier thermal processing of the resins leads to formation of reactive species such as hydroperoxides, carbonyl groups, or unsaturated bonds that can induce reactions that impair melt-flow properties or mechanical performance of polymers.

2. Please write out at lest 3 kinds of plastics additives both in English and in Chinese

[Reading Material]

Organic Stabilizers for Pipe

Common substitutes for lead-based PVC stabilizers in Europe are calcium/zinc systems, but the regulatory tide there poses a threat to the future of all heavy metals. In response, some additive producers are developing PVC pipe stabilizers that contain no heavy metals. Ciba Specialty Chemicals, Basel, Switzerland, is leading an effort to develop an organic-based PVC pipe stabilizer. In addition, Akcros Chemicals, Eccles, England, is looking at an organic PVC stabilizer as part of its broad program to find alternatives to lead. Meanwhile, in the U.S., a comparable effort is underway to replace lead in PVC wire and cable.

While plans by Ciba to commercialize the organic stabilizers have been complicated by the impending transfer of its stabilizers business to Witco, there is little doubt within the additives business that replacements for lead, and even zinc, will be needed. Finding a metal-free PVC stabilizer system "is really essential for the continued existence of the PVC industry in the longer-term".

Pfaendner, head of R&D for Ciba's vinyl additives business unit in Lampertheim, Germany, cited plans by northern European countries to ban lead-based PVC stabilizers - Denmark in 2000 and Sweden in 2005. And those nations are pushing to have the policy adopted by the European Union. Zinc is also likely to be restricted in Europe.

Nor do many Europeans favor switching from mostly lead-based stabilizers to the tin systems found in PVC pipe in North America. One reason is that lead- and tin-stabilized PVC have different flow properties and therefore require different process equipment. In addition, some tin stabilizers form black streaks if they come in contact with lead from recycled PVC.

Superior performance claimed. Ciba has a PVC stabilizer made up of organic "core" compounds, synergistic costabilizers, and added lubricants. Organic components are said to have certain chemical features - a heterocyclic structure and highly active reaction sites - that improve the color of PVC. The costabilizers act as scavengers for the hydrogen chloride released by heated PVC. As the organic system has no natural lubricity, the Ciba package also includes lubricants. While the system is free of heavy metals, some people note that the costabilizers, like those in most PVC stabilizer formulations, have small amounts of relatively innocuous metals like sodium and calcium.

To evaluate the properties of its new stabilizer package in pipes, Ciba has launched a collaborative research effort with resin maker Norsk Hydro ASA, Norway, and giant pipe producer Wavin, Netherlands. According to the partners, tests show the organic-based package to be superior to the calcium/zinc systems also vying to replace lead.

In the standard dehydrochlorination test, PVC with a concentration of 6phr of the organic stabilizer took about twice as long to reach a critical stage of polymer decomposition when heated to 180℃ as PVC with the same concentration of a calcium/zinc stabilizer. Similarly, the partners report that 110mm PVC pipes stabilized with their organic package show mechanicals - impact, fracture toughness, and heat distortion temperature - nearly the same as those of today's lead-based stabilizers.

Ciba says introduction of the organic stabilizers is likely in early 1998 for non-critical applications such as non-pressure sewage pipes (90% of Europe's PVC pipe market). Ciba is collecting toxicity data to obtain approval by Germany and other European countries to use the stabilizers in potable water pipes. Other people add that tests of the mechanical integrity of pressure pipes stabilized with the new products are also underway.

Some nations are developing calcium/zinc formulations. Some programs have already yielded experimental organic stabilizers as well as inorganics where metals are a "minor part" of composition. There is also a drive in the U.S. to find substitutes for lead stabilizers in PVC wire and cable.

Words and Expressions

alternative	[ɔːlˈtəːnətiv]	n./adj.	二中选一, 可供选择的办法
lead	[liːd]	n.	铅
streak	[striːk]	n.	条纹
commercialize	[kəˈməːʃəlaiz]	v.	使商业化, 使商品化
synergistic		adj.	协同
heterocyclic	[ˌhetərəuˈsaiklik]	adj.	杂环的
scavenger	[ˈskævindʒə]	n.	清道夫
integrity	[inˈtegriti]	n.	完整性

Unit 2 Plasticizers

Plasticizers are added to thermoplasts or elastomers to make them more flexible, improve proccessability, or allow them to be foamed. Generally, plasticizers are low-molar-mass liquids, and only seldom are they low-or high-molar-mass solids. Elastomers are mostly plasticized with mineral oils: typical rubber tires, for example, contain about 40% mineral oil. Phthalic esters dominate plasticizers for thermoplasts, and here, di(2-ethyl hexyl) phthalate ("dioctyl phthalate", DOP) is the most used. Polymeric plasticizers are only used in a relatively small number of cases. They are mostly polyesters or polyethers. High-molar-mass polyesters are used for polymer blends, but low-molar-mass polyesters are used as actual plasticizers. [1]Since the latter are produced by polycondensation, they have a broad molar mass distribution, and thus, monomer and oligomer components. High monomer fractions mean low polymer fraction, but quite high oligomer fractions. In such cases, they are called oligomeric plasticizers.

A distinction is also made between primary and secondary plasticizers. Primary plasticizers interact directly with the polymer chains, where secondary plasticizers are actually only diluents for the primary plasticizers. For this reason, secondary plasticizers are also called extenders. This, depending on the polymers, a given plasticizer can act as either a primary or a secondary plasticizer. For example, heavy oils are extenders for PVC, but primary plasticizers for elastomers.

Eighty to eighty-five percent of all plasticizers are used to produce plasticized PVC. The phthalates are preferentially used to plasticize certain polyurethanes, polyester resin, and phenolic resin. Phosphate esters are good plasticizers for melamine resins, unstaturated polyesters, phenolic resins, polyamides, and cellulose acetate. A total of about 500 different plasticizers are commercially available on the market.

Plasticizers increase the chain segment mobility by different molecular effects. [2]Polar plasticizers produce the *gauche* effect with polar polymer chains, that is, they increase the *gauche* conformation fraction at the expense of the *trans* conformations, and so reduce the mean rotational energy barrier. Acting as more or less good solvents, plasticizers dissolve helix structures and crystalline regions. In addition, chain segments become more separated on account of the dilution effect.

On the other hand, <u>solvation</u> does not increase chain mobility since a solvent sheath acts like a substituent and consequently increases the rotational energy barrier.   溶剂化作用

Because of the increased chain segment mobility, the glass transition temperatures, <u>moduli of elasticity</u>, <u>tensile strengths</u>, and <u>hardesses</u> are decreased, whereas the <u>extension at break</u> is increased. The change in these parameters can thus be used as a macroscopic measure of the effectivity of the plasticization. Of these parameters, only the glass transition temperature depends solely on the polymer chain mobility, all other parameters contain contributions from other effects. Thus, measurements on plasticization effectivity using glass transition temperatures, moduli of elasticity, tensile strengths, <u>elongations at break</u>, and hardnesses cannot yield identical results.   弹性模量/拉伸强度 硬度/断裂伸长率   断裂伸长率

[3] To increase segmental mobility, the plasticizer must be able to form a <u>thermodynamically stable</u> mixture with the polymer, that is, it must be compatible with the polymer, but solvents which are too good stiffen the chains by solvation. Thus, plasticizers must be solvents which are as poor as possible.   热力学稳定的

New Words

thermoplast	[ˈθəːməuplɑːst]	n.	热塑性塑料, 热塑性
flexible	[ˈfleksəbl]	adj.	柔性的, 柔软的
proccessability	[ˈprəusesəˈbiliti]	n.	加工性
foam	[fəum]	n.	泡沫, 泡沫塑料
		v.	发泡
phthalic	[ˈθælik]	adj.	邻苯二甲酸的
dominate	[ˈdɔmineit]	v.	支配, 占优势
ethyl	[ˈeθil, ˈiːθail]	n.	乙基
hexyl	[ˈheksil]	n.	己基
octyl	[ˈɔktəl]	n.	辛基
phthalate	[ˈθæleit]	n.	邻苯二甲酸酯（盐）
polyether	[ˌpɔliˈiːθə]	n.	聚醚, 多醚
polycondensation	[ˌpɔlikɔndenˈseiʃən]	n.	缩聚反应
oligomer	[ˈɔligəumə]	n.	低聚物
distinction	[disˈtiŋkʃən]	n.	区别, 差别
interact	[ˌintərˈækt]	v.	互相作用
diluent	[ˈdiljuənt]	adj.	稀释的
		n.	稀释剂
extender	[iksˈtendə]	n.	增容剂
preferentially	[ˌprefəˈrenʃəl]	adv.	择优地

melamine	[ˈmeləmi(ː)n]	n.	三聚氰胺
polyurethane	[ˌpɔliˈjuəriθein]	n.	聚氨酯
gauche	[gəuʃ]	adj.	左右式的，旁式的
trans-	[træns]	n.	反式
helix	[ˈhiːliks]	n.	螺旋，螺旋状物
segment	[ˈsegmənt]	n.	链段
parameter	[pəˈræmitə]	n.	参数，参量
plasticization	[ˌplæstisaiˈzeiʃən]	n.	增塑，塑化，塑炼
macroscopic	[ˌmækrəuˈskɔpik]	adj.	宏观的
elongation	[ˌiːlɔŋˈgeiʃən]	n.	伸长
thermodynamic	[ˈθəːməudaiˈnæmik]	adj.	热力学的
solvation	[sɔlˈveiʃən]	n.	溶剂化（作用）

Notes

[1] Since the latter are produced by polycondensation, they have a broad molar mass distribution, and thus, monomer and oligomer components. 因为后者由缩聚反应制成，它们的分子量分布宽，因而体系中既有单体又有低聚物组分。

[2] Polar plasticizers produce the *gauche* effect with polar polymer chains, that is, they increase the *gauche* conformation fraction at the expense of the *trans* conformations, and so reduce the mean rotational energy barrier. 极性增塑剂和极性高分子链间能产生旁式效应，即将分子的反式构象转变为旁式构象，这样就降低了平均内旋转能垒。

[3] To increase segmental mobility, the plasticizer must be able to form a thermodynamically stable mixture with the polymer, that is, it must be compatible with the polymer, but solvents which are too good stiffen the chains by solvation. 为了提高链段的运动能力，增塑剂必须能与高聚物形成热力学稳定的混合物，即增塑剂必须与高聚物相容，但良溶剂由于溶剂化作用则会使分子链变硬。

Exercises

1. Translate the following passage into Chinese

(1) Plasticizers give plastics flexibility and durability. Plasticizers hold 65 percent of the 7.5 million tons world market for plastics additives, accounting for 4.9 million tons, with a value of $ 7.6 billion. The majority of this, about ninety percent, is used for polyvinyl chloride (PVC), a polymer used in diverse applications such as coatings, plumbing, construction materials and plastic bottles. This market is likely to experience continued growth, as the market for polyvinyl chloride has grown by of eight percent a year recently.

(2) Most common type of plasticizers are chemicals called phthalates. In particular, dioctyl phthalate (DOP) is extensively used as a plasticizer for PVC. Several factors make phthalates less than ideal for use as plasticizers. They migrate to the surface of plastics, and can then evaporate or leach into the surrounding environment. This limits the usefulness of phthalate plasticizers, as they eventually migrate out of plastics entirely, and as a result the plastics become brittle.

2. Put the following words into English

增塑剂	增塑纤维	热塑性塑料	主增塑剂
添加剂	弹性	填充剂	助增塑剂

3. Please write out at least 6～8 kinds of monomers both in English and in Chinese

[Reading Material]

PVC Additives (1)

It should be noted that PVC resins, of themselves, are of no practical use. When fused they are hard, brittle compounds. Their inherent limited heat stability make any type of processing difficult if not impossible. Therefore, in order to produce a useful product other ingredients are added to the PVC resin for the purpose of: increasing flexibility, providing adequate heat stability, improving processability, imparting aesthetic appeal.

Let's consider these ingredients in some detail.

1. PLASTICIZERS

Plasticizers are low boiling liquids or low molecular weight solids that are added to resins to alter processing and physical properties. They increase resin flexibility, softness and elongation. They increase low temperature flexibility but decrease hardness. They also reduce processing, temperatures and melt viscosity in the case of calendering.

Plasticizers fall into two categories based on their solvating power and compatibility with resins.

A. Primary Plasticizers: are able to solvate resins and retain compatibility on aging. Samples of these would be:

DOP	Dioctyl phthalate
S-711	Di (*n*-hexyl; *n*-octyl; *n*-decyl) phthalate (linear)
DIDP	Di-iso decyl phthate

B. Secondary Plasticizers: are so defined because of their limited solubility and compatibility and are, therefore, used only in conjunction with primary plasticizers. The ratio of primary to secondary depends on the type and quantity of the particular plasticizers. Secondary plasticizers are used to impart special properties such as:

- low temperature flexibility:　　DMODA (di-normal octyl decyl adipate)
　　　　　　　　　　　　　　　　DOZ (di-octyl azelate)
　　　　　　　　　　　　　　　　DOA (di-octyl adipate)
- flame retardance:　　　　　　　Reofas 65 (*tri*-iso propyl phenyl phosphate)
- electrical properties:　　　　　tri-mellitates
- cost reduction:　　　　　　　　Cereclor, chlorinated paraffins

In a separate category are the polymeric plasticizers. These are long chain molecules and are made from adipic, azelaic, sebacic acids and propylene and butylene glycols. The efficiency of polymerics is poor but volatility and migration are superior. An example of a polymeric plasticizer is Paraplex G-54.

The characteristics sought in plasticizers can be summarized as follows:

(1) efficiency - This is the level or concentration needed to give a stated hardness, flexibility or modulus.

(2) the effect on low temperature flexibility.

(3) solvating power: This influences the fluxing rate of the compound at a given temperature or at a minimum fluxing temperature.

The fluxing rate relates directly to processing time.

Permanence: This relates to volatility, extraction resistance, compatibility.

2. HEAT STABILIZERS

The chief purpose of a heat stabilizer is to prevent discoloration during processing of the resin compound. Degradation begins with the evolution of Hydrogen Chloride, at about 200 ℉ Increasing sharply with time and temperature. Color changes parallel the amount of degradation running from white to yellow to brown to black. Therefore, the need for heat stabilizers.

The most effective stabilizers have been found to be:

(1) Metal soaps: Barium -cadmium solids and liquids : Mark 725, Mark 311

(2) Organo tin compounds: octyl tin mercaptide: Mark OTM

(3) Epoxies: epoxidized soya oil (G-62)

The above are most likely most effective only when used in combination (synergism).

What are some of the criteria in choosing a stabilizer system?

(1) The ability to prevent discoloration.

(2) The amount of lubrication involved. In calandering this can be of critical importance. Mark 725 has low lubricating effect while Mark 311 contributes high lubrication effect.

(3) Plate-Out - a potential side-affect of processing and has been linked to certain barium-cadium stabilizers.

(4) Compatability with the resin systems - for obvious reasons.

(5) Resistance to sunlight staining: atmospheric discoloration.

Words and expressions

brittle	[ˈbritl]	adj.	易碎的, 脆弱的
aesthetic	[iːsˈθetik]	adj.	美学的, 审美的
appeal	[əˈpiːl]	n.	要求
decyl		n.	癸基
adipate	[ˈædəˌpeit]	n.	己二酸
azelate	[ˈæzileit]	n.	壬二酸酯
adipic	[əˈdipik]	adj.	脂肪的
sebacic	[siˈbæsik]	adj.	癸二酸的
glycol	[ˈglaikɔl]	n.	乙二醇
discoloration	[disˌkʌləˈreiʃən]	n.	变色, 污点
parallel	[ˈpærəlel]	v.	相应, 平行
cadmium	[ˈkædmiəm]	n.	镉
organo	[ˈɔːgənəu, ɔːˈgænəu]	adj.	有机金属的
mercaptide	[məˈkæptaid]	n.	硫醇盐

Unit 3 Lubricants

[1]In processing high melt viscosity polymers such as polyvinyl chloride (PVC) by extrusion, milling, calendering and injection molding, the shear forces applied cause excessive frictional heat which may lead to serious thermal stability problems. Another problem in processing PVC is to assure that the polymer releases from metal surfaces of the processing equipment. To solve these problems two types of lubricants are used. [2]Lubricants which lower the melt viscosity and control frictional heat build-up are called internal lubricants while substances which promote release are called external lubricants. These materials are used in relatively small amounts since an excess will cause processing and stability problems and structural weakness in the ultimate product. [3]In the processing of polymers such as PVC discrete particles are subjected to stress and heat until there is fusion of the discrete particles and a resulting loss of particle identity. An excess amount of an external lubricant will tend to coat the individual particles and while promoting a slippage between particles will delay fusion.

The role of the internal lubricant is to reduce the internal friction within the polymeric melt, which includes reducing heat build-up when the polymer is subjected to stress. Because of the characteristic high melt viscosities of rigid PVC an internal lubricant is usually viewed as being necessary to improve flow properties. Their use will result in an economic advantage in that less work will be expended at a given set of processing conditions. In addition, improved product appearance usually results, particularly with respect to improved surface appearance. An internal lubricant will promote fusion.

Other distinguishing characteristics of internal and external lubricants are the effects they have on fusion time. Internal lubricants show no change in fusion time as the concentration of lubricant increases in the polymer system; external lubricants lengthen fusion time with increasing concentration.

Some lubricants exhibit properties of both internal and external lubricants and are identified as internal/external. [4]The degree of each type of lubricity imparted in a specific application is dependent on the type and concentration of lubricant employed, the composition of the plastic system, the type of processing equipment, and the operating parameters of the processing system.

In some instances, one encounters undesirable <u>side effects</u> in the use of lubricants, most notably in the reduction of heat stability which can lead to such major production problems as: 　副作用

<u>Thermal degradation</u> of the thermoplastic material within the <u>extruder</u> requires a halt of process operations for cleaning out. 　热降解　挤出机

Recycling of materials is limited.

The use of thermoplastic materials in light colored goods is limited.

High levels of expensive heat stabilizers may be required.

New Words

lubricant	[ˈluːbrikənt]	n.	滑润剂
extrusion	[eksˈtruːʒən]	n.	挤出
mill	[mil]	vt.	塑炼
calender	[ˈkælində]	n./vt.	压延
shear	[ʃiə]	v.	剪切
excessive	[ikˈsesiv]	adj.	过多的，额外
frictional	[ˈfrikʃənəl]	adj.	摩擦的，摩擦力的
thermal	[ˈθəːməl]	adj.	热的，热量的
stability	[stəˈbiliti]	n.	稳定性
release	[riˈliːs]	n.	剥离
		vt.	脱离，释放
lower	[ˈləuə]	vt.	降低，减弱
discrete	[disˈkriːt]	adj.	不连续的，离散的
fusion	[ˈfjuːʒən]	n.	熔化，熔解
slippage	[ˈslipidʒ]	n.	滑动，滑移
rigid	[ˈridʒid]	adj.	刚性的
concentration	[ˌkɔnsenˈtreiʃən]	n.	浓度
lubricity	[ljuːˈbrisiti]	n.	润滑
degradation	[ˌdegrəˈdeiʃən]	n.	降解
halt	[hɔːlt]	n.	停止，中断
clean out			清除，打扫干净

Notes

[1] In processing high melt viscosity polymers such as polyvinyl chloride (PVC) by extrusion, milling, calendering and injection molding, the shear forces applied cause excessive frictional heat which may lead to serious thermal stability problems. 熔体黏度高的塑料（如 PVC）在挤出、开炼、压延和注射加工过程中受到的剪切作用会产生大量的摩擦热，而产生的热量会降低塑料的热稳定性。

[2] Lubricants which lower the melt viscosity and control frictional heat build-up are called internal lubricants while substances which promote release are called external lubricants. 能降低熔体

黏度和减少摩擦生热的润滑剂称为内润滑剂，而有利于脱模的润滑剂称为外润滑剂。

[3] In the processing of polymers such as PVC discrete particles are subjected to stress and heat until there is fusion of the discrete particles and a resulting loss of particle identity. An excess amount of an external lubricant will tend to coat the individual particles and while promoting a slippage between particles will delay fusion. 在聚合物（如PVC）的加工过程中，分散的粒子受到应力和热的作用熔化成连续的熔体。大量的外润滑剂会包裹在粒子外而使粒子间相互滑移从而延迟熔化。

[4] The degree of each type of lubricity imparted in a specific application is dependent on the type and concentration of lubricant employed, the composition of the plastic system, the type of processing equipment, and the operating parameters of the processing system. 在具体应用中，每种润滑剂的润滑效果取决于润滑剂的类型和用量、配方、加工设备和加工工艺参数。

Exercises

1. Translate the following passage into Chinese

Lubricants are incorporated in plastic compounds to provide external and internal lubrication. They eliminate external friction between the polymer and the metal surface of the processing equipment; and they improve the internal flow characteristics of the polymer, adding to the wetting properties of the compounding ingredients. Many lubricant systems used in plastic compounds perform both functions. At present the most common lubricants are synthetic and natural waxes, low molecular weight polyethylene, and metallic stearates. A major application for internal lubrication is in rigid PVC where it is necessary to eliminate the high shear rate that develops in the melt during processing. Absence of lubrication will cause degradation during processing and shorten long-term durability or impair adequate service performance of the plastic product. Other plastics requiring internal lubrication are polyolefins, polystyrenes, phenolics, melamines, cellulose acetate, acrylonitrile - butadiene - styrene polymer (ABS), nylon, unsaturated polyesters and many rubber compounds.

2. Define the following technical terms in English

　　(1) heat stabilizer　　(2) plasticizer　　(3) lubricant

[Reading Material]

PVC Additives (2)

3. FILLERS

Essentially fillers are added to formulations to reduce costs, although they may offer other advantages - such as opacity, resistance to blocking, reduced plate-out, improved dry blending. On the other side, fillers can reduce tensile and tear strength, reduce elongation, cause stress whitening, reduce low temperature performance.

The most common fillers used with PVC are calcined clays, and water-ground and precipitated calcium carbonates of particle size around 3 micrometers. Other fillers are silicas and talcs.

4. LUBRICANTS

These materials are of prime importance in PVC processing. They are described below.

(1) Improve the internal flow characteristics of the compound.
(2) Reduce the tendency for the compound to stick to the process machinery.
(3) Improve the surface smoothness of the finished product.
(4) Improve heat stability by lowering internal and/or external friction.

Examples of lubricants, with which you may be familiar, are stearic acid, calcium stearate, Wax E, polyethylene AC 617.

5. PROCESSING AIDS

These may be regarded as low-melt viscosity, compatable solid plasticizers. They are added to lower processing temperature, improve roll release on calenders, reduce plate-out, promote fusion.

They are usually added at concentrations of 5.0%. The most widely used processing aids are acrylic resins of which acryloid K 120N is an example.

6. OTHER ADDITIVES

There are several other additives which we will list and comment on briefly.
(1) Impact Modifiers: These are used in rigid vinyls to improve impact resistance. These are usually acrylic or ABS polymers used at 10~15 phr levels. Examples are: Kureha BTA 111, Blendex 301.
(2) Light Stabilizers: for resistance to ultraviolet radiation. They are used in low concentrations 0.5~1.5 phr.
(3) Flame Retardants: PVC is inherently self-extinguishing. However, the plasticizers and additives are not. Therefore, flame retardants are added. The most widely known one is antimony *tri*-oxide.
(4) Anti-Static Agents
(5) Fungicides: Vinyzene BP-5
(6) Foaming Agents: Chemicals that decompose at predetermined temperatures to produce a certain volume of gas within the molten vinyl and thereby create foam.
(7) Colorants: Both pigments and dyes can be used. However, dyes, which are soluble organic substances, are used sparingly due to their tendency toward migration and extract ability. Heat resistance of colorants must be carefully evaluated.

In summary, we have seen that a vinyl compound consists of the following components: PVC resin, plasticizer, heat stabilizer, lubricant, special additive, colorants.

Words and expressions

opacity	[əuˈpæsiti]	n.	不透明性
calcine	[ˈkælsain]	v.	烧成石灰, 煅烧
acrylic	[əˈkrilik]	adj.	丙烯酸的
extinguishing	[iksˈtiŋgwiʃiŋ]		熄灭
polyolefin	[ˌpɔliˈəuləfin]	n.	聚烯烃
static	[ˈstætik]	adj.	静态的, 静电的
talc	[tælk]	n.	滑石, 云母
fungicide	[ˈfʌndʒisaid]	n.	杀菌剂
predetermine	[ˈpriːdiˈtəːmin]	v.	预定, 预先确定

PART C Resins

Unit 1 Polyethylene

Low density Polyethylene

Polyethylene originally was called polymethylene since it first was produced in the laboratory by Von Peckmann in 1898 from diazomethane. This classical product was actually a high molecular weight crystalline linear polymer of methylene (—CH_2—). Carothers produced a comparable product in 1930 by hearting $Br(CH_2)_n$ Br with sodium. Paraffin wax is, of course, a low molecular weight polymethylene, but its molecular weight is far below that of the plastic range.

The accidental or serendipitous discovery of a polyethylene that could be produced economically was based on some high pressure studies of ethylene ($H_2C=CH_2$) by Michaels at Amsterdam. This high pressure technique from a mixture of ethylene and benzaldehyde in the presence of a trace of air.

The product from a small commercial plant built in Great Britain in 1939 proved to be the best high insulating material available for coaxial cable for radar applications. Hence, many production facilities were built during World War II.

The product was originally call high-pressure polyethylene, but now, in accord with ASTM standard nomenclature, a polymer of ethylene having a density of 0.910~0.925 g/cm^3 is called low-density polyethylene (LDPE). [1]This polymer has a high volume and thus a low density because of the many branching and the degree of crystallinity increase as the reaction pressure is increased. The degree of crystallinity in commercial LDPE is 60%~70%. The opacity due to the presence of crystals is reduced by decreasing the size of the spherulites or clusters of crystals.

The origin polymer was made by the high pressure polymerization of ethylene at 475~570 °F in the presence of 0.05% oxygen as an initiator. [2]It is now customary to conduct this polymerization in the presence of a <u>peroxy initiator</u> such as <u>benzoyl peroxide</u> in an <u>autoclave</u> or <u>jacketed tube</u> at a pressure of 30000～40000 psi. (1psi.=6894.76Pa), a pressure of 9000 psi. , is considered to be the minimum usable pressure. In some instances, this polymerization is conducted in the presence of <u>benzene</u> or <u>chlorobenzene</u> and an <u>inert gas</u> may be present as a <u>chain stopper</u>. This <u>exothermic reaction</u> is <u>terminated</u> when approximately 24% of the ethylene has been converted to polymer.

过氧类引发剂/过氧化苯甲酰/反应釜/带夹套的管式反应器

苯/氯苯/惰性气体
链终止剂/放热反应终止

A very low molecular weight polyethylene is produced by the polymerization of ethylene in benzene in the presence of larger amounts of initiator at high pressures. This product has an average molecular weight of less than 10000. It is used in place of other waxes for making candles, polishes, inks, and coating.

It is customary to add <u>anti-block agents</u> to LDPE to prevent film from sticking and to reduce the tendency of molded parts to gather dust. Carbon black is often added to protect the polymer against the effect of ultraviolet light. Compatible <u>light-colored stabilizers</u> must be added when the black color is not acceptable. <u>Clay</u> and <u>ground limestone</u> have also been used as fillers. It is a customary to characterize LDPE by its <u>melt flow index</u> i.e the weight in grams that is extruded in ten minutes at 190℃.

开口剂

浅色稳定剂
陶土/重质碳酸钙

熔体流指数（速率）

The principal end use of LDPE is as film and sheet. Polyethylene film is used as a mulch barrier, as a shrink protector for pellets, and as writing paper. About 15% of this polymer is injection molded and larger quantities are extruded as <u>tubing</u> and as paper coatings. <u>Squeeze bottles</u> are produced by <u>blow molding</u> LDPE. This polymer is also used as a coating for wire and cable.

管材/挤瓶
吹塑

New Words

polymethylene	[ˌpɔliˈmeθiliːn]	n.	聚亚甲基
diazomethane	[daiˌæzəuˈmeθein]	n.	重氮甲烷
sodium	[ˈsəudjəm, -diəm]	n.	钠
paraffin	[ˈpærəfin, -fiːn]	n.	石蜡
serendipitous	[ˌserənˈdipitəs]	adj.	偶然发现的
benzaldehyde	[benˈzældiˌhaid]	n.	苯甲醛
coaxial	[kəuˈæksəl]	adj.	同轴的，共轴的
facility	[fəˈsiliti]	n.	设备，工具
nomenclature	[nəuˈmenklətʃə]	n.	命名法
spherulite	[ˈsferjuˌlait, ˈsfiə-]	n.	球晶
peroxy			过氧
benzoyl	[ˈbenzəuil]	n.	苯(甲)酰

autoclave	[ˈɔːtəukleiv]	n.	高压容器, 高压釜
benzene	[ˈbenziːn, benˈziːn]	n.	苯
chlorobenzene	[ˌklɔːrəˈbenziːn]	n.	氯苯
exothermic	[ˌeksəuˈθəːmik]	adj.	放热的
polish	[ˈpɔliʃ]	n.	光泽, 上光剂
grind (ground, ground)	[graind]	v.	磨(碎)
limestone	[ˈlaimstəun]	n.	石灰石
mulch	[mʌltʃ]	n.	覆盖, 覆盖物

Notes

[1] This polymer has a high volume and thus a low density because of the many branching and the degree of crystallinity increase as the reaction pressure is increased. 这种聚乙烯（低密度聚乙烯）的体积大密度小，因为分子中有许多支化结构，结晶度会随着反应压力的增加而增大。

[2] It is now customary to conduct this polymerization in the presence of a peroxy initiator such as benzoyl peroxide in an autoclave or jacketed tube at a pressure of 30000～40000 psi., a pressure of 9000 psi., is considered to be the minimum usable pressure. 现在这种聚合通常采用过氧类引发剂（如过氧化苯甲酰）在高压反应釜或带夹套的管式反应器中进行，压力为30000～40000磅/平方英寸，9000磅/平方英寸被认为是可用的最低压力。

Exercises

1. Translate the following into Chinese

(1) Polyethylene is an inexpensive and versatile polymer with numerous applications. Control of the molecular structure leads to low density (LDPE), linear low density (LLDPE) and high density (HDPE) products with corresponding differences in the balance of properties

(2) Polycarbonate (PC) is a clear, colorless polymer used extensively for engineering and optical applications. It is available commercially in both pellet and sheet form. Outstanding properties include impact strength and scratch resistance. The most serious deficiencies are poor weatherability and chemical resistance.

2. Write out the original words for the following abbreviations
   PE HDPE LDPE LLDPE UHMWPE BPO DCP

[Reading Material]

High Density Polyethylene

The classical polyethylene produced in the laboratory by Von Peckmann in the last half of the nineteenth century was actually a linear polymer. Since a linear polymer chain occupies less volume than a branched polymer china, it has a higher density and its structure corresponds to the product called high density polyethylene (HDPE).

The first commercially feasible process for the production of HDPE was developed by Ziegler in Germany in 1954. A comparable product was also developed in the Phillips laboratories in 1955 and in the laboratories of Standard Oil Company at a later date. The original name of low pressure polyethylene was used to describe this linear polymer, but the term high-density polyethylene is preferred.

According to ASTM standards, HDPE has a density of $0.941\sim0.965 g/m^3$. Any polyethylene having a density of $0.926\sim0.940 g/m^3$ is called medium-density polyethylene. The degree of crystallinity in the high-density product is greater than 85% for the Ziegler product and about 95% for the Phillips product. The Phillips process is favored for the product of HDPE in the USA, but the Ziegier process is widely used in other countries.

The catalyst used by Ziegler to produce his first linear polyethylene was aluminum triethyl, but the present practice is to use a mixture of titanium tatrachoride ($TiCl_4$) and aluminum triethyl in an inert atmosphere. The actual catalyst is produced from a reaction of these reactants, but the complex chemistry involved is beyond the scope of this textbook. The polymerization is conducted at about 70℃ at a pressure of about 80 psi. using low-boiling alkanes as the solvent. It is customary to quench the reaction with ethanol and to remove the catalyst residues by solvent washing.

The Phillips catalyst system consists of chromic oxide (CrO_3) supported on silica and alumina. This catalyst is first activated by heating at 250℃ and it is then suspended in cyclohexane. Ethylene is polymerized in this slurry at 100℃ and at a pressure of $100\sim500$ psi. Modifiers such as isobutene may also be added to control the reaction. The final product is dissolved and filtered to remove catalyst residues.

Metallic oxides, such as nickel oxide (Ni_2O_3) or molybdenum oxide (Mo_2O_3), on porous supports, such as charcoal or alumina, are used as catalysts in the Standard of Indiana process. These catalysts are activated by metallic hydrides such as sodium hydride (NaH) or calcium hydride (CaH_2) and the polymerization takes place at $230\sim270$℃ at a pressure of $40\sim80$ atm (1atm=101325Pa).

HDPE is extruded to form corrugated drain pipe, blow molded to produce containers ranging in size from 1 gal milk containers to 55 gal drums, injection molded to produce 5 gal shipping pails, and extruded to produce filaments. Other HDPE end products are used as pipe, toys, closures, and fuel tanks.

Polymers of ethylene with average molecular weights in the range of $1.5\sim3$ million are sometimes called ultrahigh molecular weight PE (UHMWPE). Since these polymers have a melt index of zero, they are more difficult to process. They are characterized by good ductility, good properties at low temperatures, and excellent resistance to abrasion.

The resistance of polyethylene coatings to elevated temperatures may be improved by crosslinking the polymer after applications. Exposure of polyethylene to high energy radiation, such as that produced by cobalt-60, will cause crosslinking. Compounds such as dicumyl peroxide are added to the polymer. Such compositions may be molded and crossilinked by heating at a higher temperature.

Words and Expressions

triethyl [traiˈeθəl] adj. 三乙(烷)基的
titanium [taiˈteinjəm, ti-] n. 钛

tatrachoride			四氯化物
reactant	[riːˈæktənt]	n.	反应物
alkane	[ˈælkein]	n.	烷烃
ethanol	[ˈeθənɔːl, -nəul]	n.	乙醇
chromic	[ˈkrəumik]	adj.	铬的
silica	[ˈsilikə]	n.	硅土
alumina	[əˈljuːminə]	n.	氧化铝(亦称矾土)
cyclohexane	[ˌsaikləuˈheksein]	n.	环己烷
isobutene		n.	异丁烯
filter	[ˈfiltə]	vt.	过滤
nickel	[ˈnikl]	n.	镍
molybdenum	[məˈlibdinəm]	n.	钼
charcoal	[ˈtʃɑːkəul]	n.	活性炭
hydride	[ˈhaidraɪd]	n.	氢化物
corrugated	[ˈkɔrəgeitid]	adj.	波纹的
ductility	[dʌkˈtiliti]	n.	延展性
cobalt	[kəˈbɔːlt, ˈkəubɔːlt]	n.	钴(Co)
crosslinking			交联
dicumyl peroxide			过氧化二异丙苯（DCP）

Unit 2　Polypropylene

　　Polymers of propylene were prepared in the nineteenth century by adding Friedel-Crafts catalysts such as aluminum bromide to propylene at very low temperatures. However, these sticky low molecular weight amorphous products had no commercial value. Useful products were produced in 1954 by Natta who used a Ziegler type catalyst system.

　　By polymerizing propylene at 70~80℃ at 100 atmospheres pressure using a slurry of titanium trichloride and aluminum diethyl chloride, Natta obtained a highly-crystalline, isotactic polypropylene. The methyl groups in isotactic polypropylene are all on the same side relative to the polymer chain.

　　The commercial product is obtained by removing a slurry off polymer and Ziegler-Natta catalyst in unrected propylene after 35%~70% of the monomer has been converted to the polymer. Catalyst residues are removed by washing the centrifuged product with methanolic hydrochloric acid, filtering, washing, and steam distilling. Any undesirable rubbery atactic polymer produced is soluble in hexane, and this solvent is used as a test to ascertain the extent of tacticity. The commercial product has an isotactic polymer-constant of more than 90% and a degree of crystallinity greater than 60%.

　　Polypropylene has a higher melting point and higher melt viscosity than HDPE, hence the melt flow index is determined at 230℃. The increase in volume caused by the presence of methyl groups on alternate carbon atoms yields a product with the low density of $0.9g/m^3$. Because of the presence of stereo-regular bulky methyl groups, polypropylene retains its stiffness at temperatures up to 140℃.

　　[1]As evident from its structural formula, the carbon atoms with the methyl group have only one hydrogen atom. Since the latter is located on a tertiary carbon atom, it is more readily removed under oxidizing conditions than a hydrogen atom in HDPE. Thus, polymers like polypropylene readily form free radicals and undergo degradation. Such oxidative degradation is retarded by the addition of about 0.5% each of a substituted phenolic antioxidant and an ester of a dithiocarboxylic acid, such as dilauryl dithiopropionate. Polymeric phenyl phosphates may also be used in place of the latter stabilizer. It is also customary to add substituted benzophenones or other ultraviolet light stabilizers to polypropylene.

Polypropylene is much more <u>resistant to stress cracking</u> than LDPE. It also has the ability to withstand <u>cyclic bending</u> and thus is used for <u>molded hinge</u>. Its excellent resistance to elevated temperatures can be improved by the addition of fillers such as talc, or <u>fibrous glass</u>. Polypropylene becomes brittle at low temperatures.

抗应力开裂
反复弯折
玻璃纤维

Approximately one half of all polypropylene produced is <u>injection molded</u> to form <u>automotive parts</u>, radio or television cabinets and <u>kitchenware</u>. About one third of the total production is extruded as <u>filaments</u> used for brushes, ropes, carpets, and blankets. Fibers are also produced by the <u>fibrillation of split films</u>. The latter are used as <u>floor mats</u> and as <u>synthetic grass turf</u>.

注射成型
洗车配件
厨房用具
丝
裂膜成纤/地毯
合成草皮

About one sixth of the total production of polypropylene is used as film. [2]Because of its high <u>crystallinity</u>, this film is <u>translucent</u>, but it can be made clearer than <u>cellophane</u> by orienting the crystals in the plane of the film by <u>biaxially drawing</u> the film. <u>Microporous film</u> can be produced by controlled crystallization. Pigmented polypropylene film is also used as a substitute for writing paper.

结晶性/半透明
玻璃纸（赛璐玢）
双轴拉伸/微孔膜

Pipe is produced by the extrusion of polypropylene. Large containers are made by molding or <u>blow molding</u> this polymer. Polypropylene <u>sheets</u> are <u>thermoformed</u> for the manufacture of luggage.

吹塑
压片/热成型

New Words

polypropylene	[ˌpɔliˈprəupiliːn]	n.	聚丙烯
propylene	[ˈprəupiliːn]	n.	丙烯
bromide	[ˈbrəumaid]	n.	溴化物
slurry	[ˈslə:ri]	n.	浆，泥浆，料液
titanium	[taiˈteinjəm, ti-]	n.	钛
isotactic	[ˌaisəuˈtæktik]	adj.	等规的，全同立构的
hexane	[hekˈsein]	n.	己烷
tacticity	[tækˈtisiti]	n.	立构规整性
methanol	[ˈmeθənɔl, -nəul]	n.	甲醇
bulky	[ˈbʌlki]	adj.	大的，容量大的
methyl	[ˈmeθil, ˈmiːθail]	n.	甲基
stiffness	[ˈstifnis]	n.	坚硬，硬度
tertiary	[ˈtə:ʃəri]	adj.	叔，第三的，第三位的
oxidative	[ˈɔksideitiv]	adj.	氧化的，具有氧化特性的
degradation	[ˌdegrəˈdeiʃən]	n.	降解，降级
retard	[riˈtɑ:d]	vt.	延迟，使减速，阻止
phenolic	[fiˈnɔlik]	adj.	酚的，石炭酸的
antioxidant	[ˈæntiˈɔksidənt]	n.	防老剂，抗氧化剂

dithiocarboxylic	[daiˈθaiəuˈglaikəul]	adj.	二硫代碳酸的
dilauryl	[daiˈlɔːriˌlɔ-]	n.	二月桂基
dithiopropionate		n.	二硫代丙酸酯（盐）
phosphate	[ˈfɔsfeit]	n.	磷酸盐(酯)
ultraviolet	[ˈʌltrəˈvaiəlit]	adj.	紫外线的
turf	[təːf]	n.	草根土, 草皮
mat	[mæt]	n.	垫子
translucent	[trænzˈljuːsnt]	adj.	半透明的
biaxial	[baiˈæksiəl]	adj.	双轴（的）

Notes

[1] Since the latter is located on a tertiary carbon atom, it is more readily removed under oxidizing conditions than a hydrogen atom in HDPE. 由于后者（碳原子）位于叔碳上，在氧化条件下比 HDPE 中的氢更容易被脱去。

[2] Because of its high crystallinity, this film is translucent, but it can be made clearer than cellophane by orienting the crystals in the plane of the film by biaxially drawing the film. 由于聚丙烯结晶度高用它制成的薄膜是半透明的，但可以通过对薄膜双向拉伸取向使薄膜比玻璃纸还透明。

Exercises

1. Translate the following passage into Chinese

(1) One industry source describes PP as "a least fussy resin" in terms of its willingness to mate with short- or long-glass fibers, talc, mica, and calcium carbonate particles, or the fast-expanding range of natural fibers (other than wood flour) now being explored or employed in many roles.

(2) Further, long-glass PP can be molded into parts using less complex and less costly processes than the ones used for glass-mat-reinforced PP. Potential downsides are glass-fiber orientation, part distortion, surface quality debits, and a cost about double that of short-glass PP (and equal to that of short-glass nylon.)

2. Put the following words into Chinese

aluminum bromide	molecular weight	amorphous polymer
titanium trichloride	extent of tacticity	crystalline polymer
aluminum diethyl chloride	isotactic polymer	melting point
Ziegler-Natta catalyst	atactic polymer	melt flow index
oxidative degradation	degree of crystallinity	biaxially drawing
phenolic antioxidant	resistant to stress cracking	ultraviolet light stabilizer

[Reading Material]

Nanometer Plastic

Presently, Nanometer plastic has become a popular point in the plastic science and technology. It has interested many enterprises in plastic industry, and some enterprises have put their production in the name of "Nanometer" so that to raise their level and fame even they have no conception of nanometer.

Nanometer material subject is absolutely a new science risen in recent years. It involves multidomains of knowledge such as aggregation physics, chemistry, material, biology etc. This article will only introduce some general knowledge for nanometer plastic.

Firstly we should understand nanometer, then nanometer technique and nanometer plastic last. Nanometer is a unit of length as a standard of measurement, and its dimension is 10^{-9} meter.

Generally nanometer material means that the dimension of a phase between two phases in the microstructure must be nanometer level. Nanometer granule has a very small diameter, its surface energy is very strong and it becomes aggregate very easy. So this is a very complex technical problem to get nanometer granule. Presently the nanometer granule which can be made and used is inorganic nanometer granule mostly. The nanometer granules which can modify plastic effectively are SiO_2, TiO_2, $CaCO_3$, montmorillonite (MMT) etc.

The core contents of nanometer technique is that how can solve a problem for its aggregate. It is very difficult to get single and dispersive nanometer granule as it is aggregated of itself very easy and the key of the nanometer technique is to disperse and homogenize the nanometer granule to a matrix. There are two ways that can industrialize nanometer material according to the report of application and research for nanometer technique. One is the bedded insertion technique, that means to make the base granule which has certain solid density and equal size with n-MMT processed by bedded insertion, then mix and granulate it so that can solve the problem which nanometer material is not homogeneous to disperse and get compound nanometer material; Two is to use vibratory mill dispersion to homogenize nanometer material in matrix, its aggregation is not formed in main and the dispersion is in nanometer level.

Nanometer plastic means that the matrix must be macromolecular polymer. Nanometer plastic has been improved for heat resistance, climate proof and abrasion resistance by the way that nanometer granule is dispersed in plastic resin fully. Nanometer plastic can get the rigidity and heat resistance that it is like ceramics and also has original property of plastic as tenacity and high impact resistance. Now NPE, NPET and NPA6 modified by nanometer granule can be industrialized. To put and steady silver (Ag^+) into microporousness on surface of nanometer granule can get carrying silver nanometer material which has bacteria resistance property and to put this material into plastic can make the plastic material that has some good property as bacteria resistance, mildew resistance and self-cleaning etc. and it can become green environmental protection product. Now this technique has been applied in ABS, S-PVC, HIPS, PP etc.

Nanometer plastic is a sort of new material in high technology, it has bright development prospects. Many manufacturer begin to produce nanometer plastic in a situation that not complete quality guarantee system and production management and also they are lacking in knowledge of this new material. That will engender danger that nanometer plastic comes to a premature end. So we hope the government concerned can make a complete standard and rules to protect the development of this new material.

Words and Expressions

nanometer	[ˈneinəˌmiːtə]	n.	纳米
multidomain		n.	多畴
aggregation	[ˌægriˈgeiʃən]	n.	聚集物，聚集（态）
granule	[ˈgrænjuːl]	n.	小粒，颗粒
montmorillonite	[ˌmɔntəməˈriləˌnait]	n.	蒙脱石
homogenize	[həˈmɔdʒənaiz]	v.	均质化，使均质
matrix	[ˈmeitriks]	n.	基质，基体
engender	[inˈdʒendə]	v.	造成
premature	[ˌpreməˈtjuə]	adj.	未成熟的，太早的，早熟的
microporousness		n.	微孔性

Unit 3　Polyvinyl Chloride

Polyvinyl chloride is one of the most important thermoplastic material. [1]It is finding wide application both in industrial and agriculture productions and in daily life, owing to its rich resources, low costs, good mechanical properties and corrosion resistance.

[2]Generally speaking, polyvinyl chloride is obtained by the polymerization of monomer vinyl chloride, or in more detail, it is prepared by heating water emulsion of vinyl chloride in the presence of a small amount of benzoyl peroxide as catalyst in closed vessels under pressure. Thus, monomer vinyl chloride is the foundation of polyvinyl chloride.

There are a lot of ways to obtain monomer vinyl chloride. Mainly it is obtained by reacting acetylene with hydrochloric acid gas, or by reacting ethylene obtained by cracking petroleum with chlorine to form ethane bichloride first, then cracking it to give vinyl chloride.

Polyvinyl chloride is obtained as a white fluffy powder which, when hot moulded alone, can be converted into a hard, horny mass. When it is hot masticated with 50 to100 percent of its weight of plasticizer, a profound change in properties is brought about. This plasticised PVC is rubbery and is extremely tough and flexible. PVC decomposes fairly readily when overheated, therefore temperatures must be carefully controlled during processing.

[3]Polyvinyl chloride has a succession of large negative chlorine atoms on alternate carbon atoms, which stiffen the main chain by both steric hindrance and electronic repulsion and also provide intermolecular attraction between adjacent chains. The configuration is also fairly regular, producing 5% to 10% crystallinity. Altogether these forces produce fairly rigid, strong products that have good chemical resistance and soften gradually at higher temperatures. The high crystalline melting point, however, makes for difficult processability, and the aliphatic chlorine atoms permit dehydrochlorination, which produces poor thermal stability. Polyvinyl chloride is more important in plasticized form.

[4]With variety of plasticizer and stabilizer, PVC can be made in soft, rigid or transparent form, which can be molded by injection or extrusion into many items such as raincoats, curtains, toys, and radio components.

New Words

thermoplastic	[,θə:mə'plæstik]	adj.	热塑性的
		n.	热塑性塑料
corrosion	[kə'rəuʒən]	n.	侵蚀，腐蚀状态
vessel	['vesl]	n.	容器，器皿
acetylene	[ə'setili:n]	n.	乙炔，电石气
hydrochloric	[,haidrəu'klɔ:rik]	adj.	氯化氢的，盐酸的
petroleum	[pi'trəuliəm]	n.	石油
fluffy	['flʌfi]	adj.	松散的，蓬松的
masticate	['mæstikeit]	v.	塑炼，破料
profound	[prə'faund]	adj.	深刻的，深奥的
extremely	[iks'tri:mli]	adj.	极端的，非常的
negative	['negətiv]	adj.	负的，阴性的
steric hidrance			位阻
repulsion	[ri'pʌlʃən]	n.	排斥，推斥
aliphatic	[,æli'fætik]	adj.	脂肪族的，脂肪质的
dehydrochlorination	[di:,haidrəklɔ:ri'neiʃən]	n.	脱氯化氢

Notes

[1] It is finding wide application both in industrial and agriculture productions and in daily life, owing to its rich resources, low costs, good mechanical properties and corrosion resistance. 聚氯乙烯由于来源广、价廉、力学性能和抗腐蚀性能好而广泛应用于工业、农业和日常生活中。

[2] Generally speaking, polyvinyl chloride is obtained by the polymerization of monomer vinyl chloride, or in more detail, it is prepared by heating water emulsion of vinyl chloride in the presence of a small amount of benzoyl peroxide as catalyst in closed vessels under pressure. 一般聚氯乙烯由氯乙烯单体聚合而成。详细情形是将氯乙烯水乳液中加入少量过氧化苯甲酰（BPO）作为催化剂，在密闭反应器中加压、加热进行聚合。

[3] Polyvinyl chloride has a succession of large negative chlorine atoms on alternate carbon atoms, which stiffen the main chain by both steric hidrance and electronic repulsion and also provide intermolecular attraction between adjacent chains. 聚氯乙烯分子链中每隔一个碳原子上有一个大的带负电的氯原子，这些氯原子会通过空间位阻和推电子效应使主链柔性下降并且使分子间力变大。

[4] With various of plasticizer and stabilizer PVC can be made in soft, rigid or transparent form, which can be molded by injection or extrusion into many items such as raincoats, curtains, toys, and radio components. 加入不同量的增塑剂和稳定剂，聚氯乙烯能制成或软或硬或透明的制品，采用注射或挤出成型方法制成各种制品如雨衣、窗帘、玩具和电器元件。

Exercises

1. Translate the following passage into Chinese

Emulsion polymerization consists of emulsifying very small VCM droplets in water, with a water soluble free radical catalyst . Depending on the type of "soap" or emulsifying agent, agitation, and temperature, Emulsion PVC of varying molecular weight is produced. These resin particles are much smaller than Suspension PVC, and are smooth surfaced, non absorbent to plasticizers at ambient temperatures. Emulsion PVC resins (E-PVC), also called "Dispersion resins" and "Paste resins", are used to make Plastisols and Organosols for molding, dipping and coating applications.

2. Put the following words into English

热稳定性	氯	乙烯	聚氯乙烯
空间位阻	氯乙烯	乙炔	增塑聚氯乙烯
	二氯乙烯	氯化氢	

[Reading Material]

What is PVC? (1)

PVC, Poly(vinyl chloride), or "Vinyl" is the second largest volume plastic resin produced and consumed worldwide.

PVC resin is a product of the polymerization of vinyl chloride monomer or VCM (CH_2=CHCl), in a "head-to-tail" manner via free radical catalysts. The resultant (ideal) PVC is a hydrocarbon chain (like polyethylene) but with a chlorine atom on every other carbon (—CH_2—CHCl—CH_2—CHCl—CH_2—CHCl—). Being an imperfect world, there is some chain branching during polymerization, which are weak points subject to degradation.

How is it made?

The main polymerization methods for VCM include suspension, emulsion, and bulk or mass methods. Solution polymerization, once used for coil-coating PVC's, is no longer employed.

In suspension polymerization, VCM droplets (containing free radical catalyst) are agitated with suspending agents in water for a given time and temperature to achieve the desired molecular weight. This is the most common production method, and furnishes "popcorn-like", irregularly shaped resin grains that can absorb liquid plasticizers and additives to form dryblend powder compounds. Most flexible and rigid PVC calendering, molding, and extrusion (from powder or pellets) is done with Suspension PVC (S-PVC).

Emulsion polymerization consists of emulsifying very small VCM droplets in water, with a water soluble free radical catalyst. Depending on the type of "soap" or emulsifying agent, agitation, and temperature, Emulsion PVC of varying molecular weight is produced. These resin particles are much smaller than Suspension PVC, and are smooth surfaced, non absorbent to plasticizers at ambient temperatures. Emulsion PVC resins (E-PVC), also called "Dispersion resins" and "Paste resins", are used to make Plastisols and Organosols for molding, dipping and coating applications.

Bulk (or mass) polymerization entails just the VCM monomer, containing catalyst, in a two stage reactor. The first stage reactor, with reflux condenser, agitates the VCM monomer to about a

10% conversion to polymer. This slurry is then transferred to a horizontal reactor with a ribbon blending type low RPM agitator, where polymerization is finished as a dry powder. This PVC (M-PVC) is similar in particle size and shape to S-PVC, and is used in the same (mostly rigid) processes as S-PVC. The main difference between M-PVC and S-PVC of the same molecular weight is the higher bulk density of M-PVC.

After all the above reactions are incomplete, the PVC resin is "steam-stripped" and dried in order to remove any residual VCM monomer.

Up to now, we have only discussed PVC homopolymers. With both S-PVC and E-PVC methods, vinyl chloride monomer can and is copolmerized with other co-monomers, mainly vinyl acetate, to form PVC/PVAc copolymers. For equivalent molecular weights, vinyl copolymers show lower melt viscosities, higher tolerance for additive fillers, higher burn sensitivity to Zinc-containing stabilizers, and better cold-draw properties than homopolymers. They have found some specialty application niches, to be discussed later.

Words and Expressions

popcorn-like	['pɔpkɔ:n]		米花状
dryblend		n.	干混料, 干混合
ambient	['æmbiənt]	n.	环境温度, 室温
calender	['kælində]	n.	压延机
plastisol	['plæstisɔ:l]	n.	增塑溶胶, 增塑糊
organosol	[ɔ:'gænəsɔl]	n.	稀释增塑糊, 有机溶胶
reflux	['ri:flʌks]	n.	回流
steam-strip			汽提
vinyl acetate			乙酸乙烯酯

Unit 4 Polystyrene

The family of styrene polymers includes polystyrene, copolymers of styrene with other vinyl monomers, polymers of derivatives of styrene, and mixtures of polystyrene and styrene-containing copolymers with elastomers. In comparison to the other major high-volume resins, the outlook for polystyrene is relatively unpromising. The major application field for polystyrene is packing, accounting for a thud of its end markets.

Polystyrene is a thermoplastic with many desirable properties. It is clear, transparent, easily colored, and easily fabricated. It has reasonably good mechanical and thermal properties, but is slightly brittle and softens below 100℃.

[1]Although solution or emulsion polymerization may occasionally be used, most polystyrene is made either by suspension polymerization or by polymerization in bulk. All but the latter process are typically carried out.

Polystyrene is linear polymer, the commercial product being atactic and therefore amorphous. [2]Isotactic polystyrene can be produced, but offers little advantage in properties except between the glass transition (about 80℃) and its crystalline melting point (about 240℃), where it is much like other crystalline plastics. Isotactic polystyrene is not of commercial interest because of increased brittleness and more difficult processing than the atactic product.

Like most polymers, polystyrene is relatively inert chemically. It is quite resistant to alkalis, halide acids, and oxidizing and reducing agents. It can be nitrated by fuming nitric acid, and sulfonated by concentrated sulfuric acid at 100℃ to a water-soluble resin. Chlorine and bromine are substituted on both the ring and the chain at elevated temperatures. Polystyrene degrades at elevated temperatures to a mixture of low-molecular-weight compounds about half of which is styrene. The characteristic odor of the monomer serves as an identification for the polymer.

As made, polystyrene is outstandingly easy to process. Its stability and flow under injection-molding conditions make it an ideal polymer for this technique. Its optical properties - color, clarity, and the like - are excellent, and its high refractive index (1.60) makes it useful for plastic optical components. Polystyrene is a good electrical insulator and has a low dielectric loss factor at moderate frequencies. Its tensile strength reaches about 8000 psi. 介电损耗因子/拉伸强度

On the other hand, polystyrene is readily attacked by a large variety of solvents, including dry-cleaning agents. Its stability to outdoor weathering is poor, and it turns yellow and crazes on exposure. 干洗剂 银纹

Many of these defects can be overcome by proper formulating, or by copolymerization and blending. [3]For example, the addition of ultraviolet-light absorbers improves the light stability of polystyrene enough to make it useful in lighting fixtures such as fluorescent-light diffusers. Flame-retardant polystyrene have been developed through the use of additives. 配合 共聚/共混 紫外光吸收剂 照明设备/荧光扩散器 阻燃 添加剂

New Words

polystyrene	[ˌpɒliˈstaiəriːn]	n.	聚苯乙烯
derivative	[diˈrivətiv]	adj.	衍生物
thud	[θʌd]	n.	碰击声，重击
fabricate	[ˈfæbrikeit]	v.	制作，构成
halide	[ˈhælaid]	n./adj.	卤化物，卤化物的
bromine	[ˈbrəumiːn]	n.	溴
nitrate	[ˈnaitreit]	v.	硝化
sulfonate	[ˈsʌlfəˌneit]	v.	磺化
odor	[ˈəudə]	n.	气味
optical	[ˈɔptikəl]	adj.	光学的
tensile	[ˈtensail]	adj.	可伸长的，可拉长的
agent	[ˈeidʒənt]	n.	介质，剂
fluorescent	[fluəˈresənt]	adj.	荧光的
flame-retardant		n.	阻燃剂
formulating	[ˈfɔːmjuleitiŋ]	n.	配制，配料

Notes

[1] Although solution or emulsion polymerization may occasionally be used, most polystyrene is made either by suspension polymerization or by polymerization in bulk. 虽然有时也采用溶液或乳液聚合，但大部分聚苯乙烯由悬浮聚合或本体聚合制得。

[2] Isotactic polystyrene can be produced, but offers little advantage in properties except between the glass transition (about 80℃) and its crystalline melting point (about 240℃), where it is much like other crystalline plastics. 可以合成出全同聚苯乙烯，但全同聚苯乙烯除了在玻璃化温度（约80℃）和熔点（约240℃）之间类似于结晶塑料外没有其他优点。

[3] For example, the addition of ultraviolet-light absorbers improves the light stability of polystyrene enough to make it useful in lighting fixtures such as fluorescent-light diffusers. 例如，加入紫外光吸收剂可提高聚苯乙烯的光稳定性，使之可应用于照明设备（例如荧光扩

散器)。

Exercises

1. Translate the following passage into Chinese

Polystyrene is an inexpensive and hard plastic, and probably only polyethylene is more common in your everyday life. The outside housing of the computer you are using now is probably made of polystyrene. Model cars and airplanes are made from polystyrene, and it also is made in the form of foam packaging and insulation. Clear plastic drinking cups are made of polystyrene. So are a lot of the molded parts on the inside of your car, like the radio knobs. Polystyrene is also used in toys, and the housings of things like hairdryers, computers, and kitchen appliances.

2. Translate the following sentences in the text

(1) The family of styrene polymers includes polystyrene, copolymers of styrene with other vinyl monomers, polymers of derivatives of styrene, and mixtures of polystyrene and styrene-containing copolymers with elastomers.

(2) Polystyrene is a thermoplastic with many desirable properties: clear, transparent, easily colored, easily fabricated and have good mechanical and thermal properties.

(3) Polystyrene is linear polymer, the commercial product being atactic and therefore amorphous.

(4) Like most polymers, polystyrene is relatively inert chemically. It is quite resistant to alkalis, halide acids, and oxidizing and reducing agents.

(5) Polystyrene is outstandingly easy to process. Its stability and flow under injection-molding conditions make it an ideal polymer for this technique.

(6) polystyrene is readily attacked by a large variety of solvents, including dry-cleaning agents.

[Reading Material]

What is PVC? (2)

Inherent Properties of PVC

Containing 56.5% chlorine and 43.5% ethylene from petroleum feedstocks, PVC is much less dependent than most other thermoplastic resins on the fluctuations of supply and demand of the petroleum industry. Its chlorine content is derived from table salt! PVC's chlorine content provides inherent flame & fire retardancy. Other additives (plasticizers, modifying resins) may burn, but PVC will not support combustion on its own.

PVC is regarded as perhaps the most versatile thermoplastic resin, due to its ability to accept an extremely wide variety of additives: plasticizers, stabilizers, fillers, process aids, impact modifiers, lubricants, foaming agents, biocides, pigments, reinforcements. Indeed, PVC by itself cannot be processed! It must have at least a stabilizer, a lubricant, and if flexible, a plasticizer present.

PVC products can run the gamut from a wiggly fishing worm to a high impact computer housing, pipe, windows and fencing, and all in between. Clear or opaque, flexible PVC applications (flooring, automotive, wire & cable) dominated the earlier years (40's, 50's, early 60's), but with the advent of reciprocating screw injection molding and twin screw extrusion in the 60's, rigid PVC began to flex its muscle in pipe fittings, siding, electrical junction boxes, fencing, docking, to the point today where rigid PVC applications account for about 70% of all PVC processed!

Physical properties of course will vary widely depending on types and amounts of additives chosen. Based on cost/performance, many consider rigid PVC to be "the poor man's engineering resin"!

PVC has a unique degradation sequence. Unlike most other polymers that exhibit mainly oxidative degradation with peroxide formation and chain scission, protected by antioxidants, PVC (while also undergoing oxidative degradation) has a nasty habit of releasing HCl under heat and shear of processing - an "unzippering effect" that rapidly progresses to catastrophic charred blackening if left unchecked. The art and science of stabilization - a whole industry sector - has developed very effective protective stabilizer additives to retard this type of degradation. This HCl elimination is most likely to start at a "weak link" site - typically a chlorine on a carbon at a branching site in the chain. The result is a series of alternating (or conjugated) double bonds, and the onset of visible discoloration (yellowing) has been pegged at 7～8 conjugated double bonds. However a UV black light can see early degradation at 3～4 double bonds before it becomes visible to the eye.

Words and Expressions

fluctuation	[ˌflʌktjuˈeiʃən]	n.	波动, 起伏
modifing resin			改性树脂
splinter	[ˈsplintə]	n./v.	裂片, 裂成碎片
biocide	[ˈbaiəsaid]	adj.	生物杀灭剂
unzippering effect			拉链式解聚合
chain scission	[ˈsiʒən, ˈsiʃən]	n.	断链
antioxidant	[ˈæntiˈɔksidənt]	n.	抗氧化剂
undergoing		vt.	经历, 遭受, 忍受
unzippering	[ʌnˈzipə(r)]	vt.	拉开…上的拉链
catastrophic	[ˌkætəˈstrɔfik]	adj.	悲惨的, 灾难的

Unit 5 Phenolics

That phenol (C_6H_5OH) reacts readily with aldehydes, such as formaldehyde (H_2CO), under both acid and alkaline has been known for over a century. Unfortunately, the early organic chemists, used more than 1 mole of formaldehyde per mole of phenol and obtained products that differed from the traditional crystalline products of organic chemistry reactions. [1]These chemists described these resinous masses as goop gunk, and were quick to abandon the investigation of these reactions and to return to the syntheses of more classical products.

Later attempts by more practical chemists, were more successful since they used lower ratios of formaldehyde to phenol. [2]However, the first commercial phenol formaldehyde plastic (PF) was not made until Leo Backeland produced a reproducible thermoplastic that he subsequently converted to a thermosetting plastic by heating the thermoplastic under pressure. The first commercial product, which was produced in the presence of alkaline such as ammonia or sodium carbonate, was patented by Backeland in 1909. This so-called resole resin was made with an excess of formaldehyde as a 40% aqueous solution. The change of the resin from a thermoplastic or A-stage resin to a cross-linked C-stage resin was controlled by a gradual increase in temperature. It was customary to use 1.5 moles of formaldehyde and 1 mole of phenol.

Phenolic molding powders are produced by blending the powdered A-stage novolac resin with wood flour or other fillers, pigments, mold lubricants, and a source of additional formaldehyde. The formaldehyde needed to react with the unreacted centers in the trifunctional phenol is supplied by the addition of hexamethylenetetramine. The latter is obtained by the reaction of 6 moles of formaldehyde and 4 moles off ammonia. The type of reaction taking place when this mixture is heated in a mold.

In actually practice the blend of fusible resin and hexamthylenetetramine is heated on hot differential rolls or in an extruder in order to advance the A-stage resin to the B-stage. The mixture containing this B-stage resin is cooled and pulverized. It is customary to blend this molding powder with other lots of phenolic molding powder to assure uniformity in a large quantity of the molding powder. The presence of fine particles in phenolic molding wood powders may be eliminated by pelletizing the powder.

  Wood flour is produced by attrition-type grinding waste. It is used as a filler for general-purpose molding powder. The impact resistance is improved by substituting cotton flock, fibrous glass or asbestos for the wood flour. Asbestos filler contributes to the heat resistance of the end product. They improved by the incorporation of butadiene-acrylonitrile copolymer rubber with the ground A-stage resin before the blend is advanced to the B-stage.

  Phenolic resins were the first all-synthetic polymers produced commercially. It is of interest to note that they are still being produced using recipes similar to those used by Backeland seventy years after the pioneer production of these resins.

  The resoles are used for laminates and as impregnates for wood and other porous material. [3]The molded novolac plastics are used for automotive parts, such as distributor caps, for household appliances, such as pot handles, and for electrical parts, such as wall plugs and circuit breakers.

  Specialized phenolic resins, which have better flow properties in the mold, have been produced by substituting furfural for formaldehyde. Presumably, this liquid aldehyde, obtained by heating corn cobs with acid, reacts with phenol in a manner similar to that of formaldehyde. Cresols, which are methylphenols, are also used in place of phenol. A thermosetting plastic can not be produced from cresols if the methyl group is present in the 2 or 4 positions on the benzene ring.

造粒

抗冲击

石棉/耐热性
丁腈橡胶

配方

层压板/浸渍
线性酚醛树脂
汽车配件/配电器盖
锅把/电子器件/墙上插座
断路器
流动性
糠醛
玉米棒
甲酚/甲酚
苯环

New Words

aldehyde	[ˈældihaid]	n.	醛，乙醛
formaldehyde	[fɔːˈmældiˌhaid]	n.	甲醛，蚁醛
alkaline	[ˈælkəlain]	adj.	碱性的，碱的
gunk	[gʌŋk]	n.	黏性物质，无特殊形状之一堆
resole	[ˈriːsəul]	n.	体型酚醛树脂
novolac		n.	线型酚醛树脂
hexamethylenetetramine		n.	六亚甲基四胺
pulverize	[ˈpʌlvəraiz]	v.	研磨成粉
laminate	[ˈlæmineit]	n.	层压板
impregnate	[ˈimpregneit]	n.	浸渍
furfural	[ˈfəːfərəl]	n.	糠醛
cresol	[ˈkriːsɔl]	n.	甲酚

Notes

[1] These chemists described these resinous masses as goop gunk, and were quick to abandon the investigation of these reactions and to return to the syntheses of more classical products. 化学家形容这些树脂是黏糊糊的废弃物，很快放弃对这些反应的研究转而进行较为传统产品的合成。

[2] However, the first commercial phenol formaldehyde plastic (PF) was not made until Leo Backeland produced a reproducible thermoplastic that he subsequently converted to a thermosetting plastic by heating the thermoplastic under pressure. 然而，直到 Leo Backeland 制成了可再加工的热塑性树脂才出现了最早的商品化酚醛树脂，他接着在一定压力下加热这种热塑性树脂将其转化成热固性树脂。

[3] The molded novolac plastics are used for automotive parts, such as distributor caps, for household appliances, such as pot handles, and for electrical parts, such as wall plugs and circuit breakers. 模塑酚醛树脂应用于制作汽车配件（如配电器盖）、家用器具（如锅把）和电器（如插座和断路器）。

Exercises

1. Translate the following passage into Chinese

Poly(methyl methacrylate) (PMMA) is a clear, colorless polymer used extensively for optical applications. It is available commercially in both pellet and sheet form. Outstanding properties include weatherability and scratch resistance. The most serious deficiencies are low impact strength and poor chemical resistance.

2. Put the following words into English

苯酚	甲醛	酚醛树脂	酸
甲酚	糠醛	甲阶树脂	碱
碳酸钠	碳酸铵	六亚甲基四胺	盐

[Reading Material]

Future 15 Years in Prospect of Chemical Construction Material

Chemical construction material will be faced an enormous market, Vigorously developing and popularizing, it will bring about outstanding economic and social benefit. The government will further strengthen to support and develop chemical construction material, emphasize to apply plastic pipe, plastic door and window, new waterproof material and construction paint and spur on developing other chemical construction material at the same time.

There are about 2000 production lines for plastic pipe and 2100 production lines for plastic door and window in China presently. The total output was 450000 Tons for plastic pipe and fittings and 200000 Tons for plastic door and window in 1998.

Up to 2010, the plastic pipe will cover 50% market of all sorts of pipe and the plastic door and window will cover more than 20% market of all sorts of construction door and window in the whole country. The plastic pipe made of PVC-U and PE will be applied, and other new plastic pipe will be developed. In the whole country, 70% construction drainage, 60% drinking water pipe and hot water pipe, 50% construction rain pipe and 80% wire preservation pipe will be plastic in all building projects. The plastic pipe will be used for public water main (under DN 400mm) to 50% in city and 70% in countryside. The plastic pipe also will be used for 30% of urban drainage and 20% of urban gas main (middle and low pressure). Plastic door and window made of UPVC has been popularized mainly. Its market rate will reach 20% in all construction door and window, and the rate of use plastic door and window will be more than 50% in heating district and not lower than 35% in other district needed saving energy in construction. Therefore, about 40000000 square metres plastic door and window will be needed by 2010.

The development to 2015: 80% construction drainage will be plastic pipe and the cast iron pipe will be obsolete mostly; 90% wire preservation pipe will be plastic pipe; 80% construction rain pipe will be plastic pipe; 85% construction drinking pipe, hot water pipe and heating pipe will be plastic pipe and the zinc coating pipe will be obsolete mostly; The public water main which is in 80% in city and 90% in countryside will be plastic pipe; more than 40% of urban gas main and 50% urban drainage will be plastic pipe. The rate of using plastic door and window will reach 60% in the district needed saving energy, the rate of market will reach 40%. And 80000000 square metres plastic door and window will be needed in the whole country.

Words and Expressions

construction material			结构材料
vigorously	['vigərəsli]	adv.	精神旺盛地
outstanding	[aut'stændiŋ]	adj.	突出的, 显著的
social benefit			社会利益
strengthen	['streŋθən]	v.	加强, 巩固
emphasize	['emfəsaiz]	vt.	强调, 着重
waterproof	['wɔ:təpru:f]	adj.	防水的, 不透水的
spur	[spə:]	v.	鞭策, 刺激, 驱策
paint	[peint]	n.	油漆, 颜料, 涂料
drainage	['dreinidʒ]	n.	排水装置
preservation	[,prezə(:)'veiʃən]	n.	保存
UPVC		n.	未增塑聚氯乙烯
obsolete	['ɔbsəli:t]	adj./n.	陈旧的, 荒废的

Unit 6　Epoxy Resins

The "ep" in the word "epoxy" is derived from the Greek prefix meaning over or between, and the "oxy" is of course, the English abbreviation for oxygen.

When bisphenol A is condensed with epichlorohydrin in the presence of alkali, the resulting diglycidyl ethers are called epoxy resins. By using excess epichlorohydrin and minimum alkali to control molecular weight, these prepolymers are made to be viscous liquids or soluble or spread on surfaces as coatings or adhesives. When mixed with polyamines, they cure rapidly at room temperature. [1]When mixed with polybasic acid anhydrides, they can be cured at higher temperatures to give harder and more heat-resistant properties.

There is a tremendous range of epoxy curing agents available today Some of the more important types are listed below.

Aliphatic Amines:

Triethylene tetramine, diethylene triamine, ethylene diamine, hexamethylene diamine. The primary amine group can react with two epoxy groups thus promoting chain formation. The secondary amine group reacts only once, and tertiary amine groups do not condense, but as an accelerator. Aliphatic amines are widely used for room temperature curing applications.

Aromatic Amines:

Meta-phenylene diamine, methylene dianiline. Aromatic amines react slower and generally require heat to cure completely. The cured resins have some what higher heat distortion temperature than those obtained with aliphatic amines.

Carboxylic Acid anhydrides:

Phthalic anhydride, hexahydrophthalic anhydride, dodecenylsuccinic anhydride. This is a large group which offers good properties but requires prolonged heating to achieve full cure.

When the plastic is cast around delicate electrical and electronic equipment, this is referred to as potting or encapsulation. [2]When used for coatings and adhesives, the reactivity of the substrate helps to provide high adhesion, and the high crosslinking provides heat and chemical resistance.

New Words

epoxy	[eˈpɔksi]	n.	环氧树脂
abbreviation	[əˌbriːviˈeiʃən]	n.	缩写
bisphenol A		n.	双酚 A
epichlorohydrin	[ˌepiˌklɔːrəˈhaidrin]	n.	环氧氯丙烷
diglycidyl		n.	缩水甘油基
ether	[ˈiːθə]	n.	醚
prepolymer		n.	预聚物
polyamine	[ˌpɔliəˈmiːn]	n.	多元胺
cure	[kjuə]	v./n.	固化
polybasic	[ˌpɔliˈbeisik]	adj.	多元的
anhydride	[ænˈhaidraid]	n.	酐
aliphatic	[ˌæliˈfætik]	adj.	脂肪族的
triethyl	[traiˈeθəl]	adj.	三乙基的
tetramine	[tetrəˈæmin]	n.	四胺
triamine	[traiˈæmin]	n.	三胺
diamine	[daiˈæmiːn]	n.	二胺
hexamethylene	[ˌheksəˈmeθiliːn]	n.	己基
aromatic	[ˌærəuˈmætik]	adj.	芳香的
methylene	[ˈmeθiliːn, -lin]	n.	亚甲基
dianiline	[daiˈæniliːn]	n.	二苯胺
carboxylic	[ˌkɑːbɔkˈsilik]	adj.	羧基的
phthalic	[ˈθælik]	adj.	邻苯二甲酸的
hexahydrophthalic	[ˌheksəˌhaidrəˈθælik]	adj.	六氢化邻苯二甲酸的
dodecenylsuccinic		adj.	十二碳烯基琥珀酸的
encapsulation	[inˌkæpsjuˈleiʃən]	n.	包胶, 封装

Notes

[1] When mixed with polybasic acid anhydrides, they can be cured at higher temperatures to give harder and more heat-resistant properties. 环氧树脂和多元酸酐混合后可以在较高温度下固化，并且硬度更大和耐热性提高。

[2] When used for coatings and adhesives, the reactivity of the substrate helps to provide high adhesion, and the high crosslinking provides heat and chemical resistance. 用作涂料和胶黏剂时，环氧树脂与基体间的反应使粘接性能提高，同时交联度提高可使耐热性和耐化学药品性提高。

Exercises

1.Translate the following passage into Chinese

Research into mechanisms of mixing is challenging due to the complex flow fields, high temperature gradients, multiphase flow, and non-Newtonian rheologies encountered during polymer processing. In addition, commercial practice on large extruders operating at high speeds makes the material being processed relatively inaccessible to the investigator. Fortunately, laboratory batch intensive mixers can often be used to simulate more complex processing operations.

2. Put the following words into English

热塑性塑料	热固性塑料	聚乙烯	聚丙烯
聚氯乙烯	聚丙烯	聚苯乙烯	聚氯乙烯
酚醛树脂	环氧树脂	醚	醇
酚	羧酸	胺	酸酐

[Reading Material]

Plastics and the Environment

Issues affecting plastics and their relationship to the environment are becoming more integral to everyday business concerns. Processors, and the industry generally, now deal with environmental goals pertaining to material selection, design, recycle content, and degradability as a matter of course. Increasingly, environmental considerations are being reconciled with real-world needs, notably, quality, end-product value, and profit-making.

The debate about plastics' role in the environment is becoming more pragmatic. In the late 1980s, the industry worldwide became the victim of numerous myths. Critics charged that plastics were major waste contributors; were harder to recycle than other material; were unsafe to burn; and constituted a drain on the earth's resources. They demanded that plastics be made to "disappear" (i.e., degrade) or else to be recycled in closed-loop systems.

Once considered by many to be environmentally suspect, plastics today are on an even footing in intermaterial competition. This report details notable examples of progress being made toward meeting environmental goals.

Cradle-to-grave environmental assessments favor plastics' use in many markets. Life-cycle assessment is a comprehensive means of comparing the impact that different material, product designs, fabrication processes, and waste-disposal methods have on the environment. This emerging technique could cast plastics use in a more favorable light versus competitive Material.

Life-cycle assessment (LCA) is a "cradle-to-grave" approach for evaluating a product throughout its life - from fabrication to disposal. Many observers expect the use of LCA to level the playing field for plastics in intermaterial competition, because it provides a broader perspective of a product's environmental impact than just its recyclability.

Precision degradables redefine application roles .The next generation of biodegradable polymers has an assured but highly specialized niche in world markets. These emerging Material offer more controlled degradation and provide a greater range of end-use benefits.

Degradables are no longer seen as a panacea for disposing of waste, degradability is seen as another form of recycling. In the market it is considered most viable when the reclaim value of a product is zero, or when overall cradle-to-grave environmental benefits favor degradability.

Moreover, improved grades and increased supplies of such degradable resins as polylactic acid (PLA), polyvinyl alcohol (PVA), polyester, cellulosics, and other biopolymers are becoming more widely available.

Environmentally optimized packages stimulate demand. Environmental criteria are progressively being integrated into plastic food packaging. Designers are incorporating features that reduce or avoid waste and pollution effects at the source, or foster recycle-use. Environmental goals now coexist equally with conventional parameters. Indeed, an emerging consensus favoring the "environmentally-optimized package" could well prove a boon for plastics processors in decades ahead.

Words and Expressions

integral	['intigrəl]	adj.	构成整体所必需的
degradability	[di͵greidə'biliti]	n.	降解性
pragmatic	[præg'mætik]	adj.	实际的，注重实效的
recyclability	['riː'saikl]	n.	再循环性，再回用性
polylactic acid		n.	聚 2-羟基丙酸, 聚乳酸
polyvinyl alcohol	[͵pɔli'vainil]	n.	聚乙烯醇
streamlined	['striːmlaind]	adj.	最新型的，改进的
biodegradable	[͵baiəudi'greidəbl]	adj.	生物所能分解的
cellulosics	[selju'ləusiːn]	n.	纤维素
criteria	[krai'tiəriə]	n.	标准
integrate	['intigreit]	vt.	使成整体，结合
foster	['fɔstə]	vt.	鼓励
consensus	[kən'sensəs]	n.	一致同意，多数人的意见
optimize	['ɔptimaiz]	vt.	使最优化

PART D Rubber Compounding Ingredients

Unit 1 General Introduction

绪论

Rubber plays a vital role in all our lives and has become essential for our comfort, safety and prosperity. It is extensively used in transport, construction, industry, healthcare and the home.

[1] Rubber is a complex material and it has to be mixed with several other <u>ingredients</u>, not only to <u>achieve the performance levels</u> of the <u>finished product</u> but also to <u>facilitate manufacture</u> and <u>reduce costs</u>. The key additives in a <u>rubber formulation</u> are as follows.

配合剂/达到性能指标
成品/便于加工/降低成本
橡胶配方

Rubber - this may be natural rubber, one of over 30 types of <u>synthetic rubber</u> or a <u>blend of rubbers</u>.

合成胶/并用胶

[2] <u>**Vulcanizing System**</u> - this is a combination of ingredients necessary to convert the rubber from a plastic-like mass to a fully elastic material through the process of <u>vulcanization</u> in which polymer molecules become <u>crosslinked</u>. The most commonly used <u>vulcanizing agent</u> is <u>sulphur</u>, with the reaction time shortened by adding one or more <u>accelerators</u>.

硫化体系

硫化
交联/硫化剂/硫黄
促进剂

<u>**Fillers**</u> - these are normally particulate materials, typically <u>carbon black</u>, <u>silica</u>, <u>clay</u> and <u>calcium carbonate</u>, and are added to reinforce stiffen, cheapen and aid processing.

填料/炭黑
白炭黑/陶土/碳酸钙

Oils - these are added to soften, plasticize, reduce costs and facilitate processing.

[3] <u>**Antidegradants**</u> - these are added in relatively small quantities to <u>guard against deterioration</u>, the most commonly used types being <u>antioxidants</u> to inhibit <u>oxidative ageing</u>, <u>antiozonants</u> to arrest ozone attack and <u>anti-flex cracking agents</u> to enhance <u>resistance to fatigue cracking</u>.

防老剂
防止老化
抗氧剂/氧老化/抗臭氧剂
抗屈挠龟裂剂
抗疲劳龟裂

Other **special purpose ingredients** include <u>blowing agents</u> (for <u>cellular rubber production</u>), <u>pigments</u>, <u>fire retardants</u> and a range of <u>processing aids</u>.

专用配合剂/发泡剂
海绵胶/颜料/阻燃剂
加工助剂

Some **formulations** will contain as many as 20 separate <u>compounding ingredients</u> and many manufacturers may employ several hundred formulations, thus necessitating the use and storage of a large number of ingredients.

配合剂

New Words

prosperity	[prɔsˈperiti]	n.	繁荣
performance	[pəˈfɔːməns]	n.	性能
blend	[blend]	vt.	混合，并用
		n.	并用胶，混合胶，
vulcanize	[ˈvʌlkənaɪz]	v.	硫化
convert	[kənˈvəːt]	vt.	使转变，转换
elastic	[iˈlæstik]	adj.	弹性的
process	[prəˈses]	n.	过程，作用，程序
		vt.	加工，操作
sulphur	[ˈsʌlfə]	n.	硫黄
accelerator	[ækˈseləreitə]	n.	促进剂
particulate	[pəˈtikjulit, -leit]	n.	微粒
		adj.	微粒的
silica	[ˈsilikə]	n.	白炭黑
clay	[klei]	n.	黏土，陶土
stiffen	[ˈstifn]	vt.	刚性，使变硬
soften	[ˈsɔ(ː)fn]	v.	(使)变柔软
plasticize	[ˈplæstisaiz]	v.	增塑
antidegradant		n.	防老剂，抗降解剂
deterioration	[di,tiəriəˈreiʃən]	n.	变坏，退化
antioxidant	[ˈæntiˈɔksidənt]	n.	抗氧剂，防老剂
oxidative	[ˈɔksideitiv]	adj.	氧化的
antiozonant	[ˈæntiˈəuzənent]	n.	抗臭氧剂
ozone	[ˈəuzəun, əuˈz-]	n.	臭氧
flex	[fleks]	vt. / n.	曲挠，多次弯曲
cracking	[ˈkrækiŋ]	n.	龟裂，裂口，裂纹
retardant	[riˈtɑːdənt]	n.	延缓(作用)剂
formulation	[,fɔːmjuˈleiʃən]	n.	配方，配料，配合
separate	[ˈsepəreit]	adj.	个别的，单独的，单个的
manufacturer	[,mænjuˈfæktʃərə]	n.	制造业者，厂商
necessitate	[niˈsesiteit]	v.	成为必要

Notes

[1] Rubber is a complex material and it has to be mixed with several other ingredients, not only to achieve the performance levels of the finished product but also to facilitate manufacture and reduce costs. 橡胶是各种物料混合而成的材料，生胶中必须要加入许多配合剂进行混合，这样不仅能使成品达到其技术指标，而且有利于加工和降低成本。

[2] Vulcanizing System - this is a combination of ingredients necessary to convert the rubber from a

plastic-like mass to a fully elastic material through the process of vulcanization in which polymer molecules become crosslinked. 硫化体系是由几种配合剂组合而成的体系，它能通过硫化工艺过程使得橡胶由塑性转变成完全弹性的材料，在硫化过程中橡胶分子变得交联了。

[3] Antidegradants - these are added in relatively small quantities to guard against deterioration, the most commonly used types being antioxidants to inhibit oxidative ageing, antiozonants to arrest ozone attack and anti-flex cracking agents to enhance resistance to fatigue cracking. 防老剂是在胶料中的加入量较少，能抑制胶料老化的助剂，最常用的防老剂有防止氧老化的抗氧化剂，防止臭氧老化的抗臭氧剂和防止疲劳老化的抗屈挠龟裂剂。

Exercises

1. Read and comprenhese the following recipe

Typical Tyre Tread Recipes

Ingredient	phr		Function
	NR	SR	
SS(Smoked sheet)	100	—	elastomer
SBR/oil masterbatch	—	103.1	elastomer-extender masterbatch
cis-BR	—	25	general purpose elastomer
Oil soluble sulfonic acid	2.0	5.0	processing aid
SA	2.5	2.0	accelerator-activator
ZnO	3.5	3.0	accelerator-activator
PBN	2.0	2.0	antioxidant
Substituted N, N'- p - phenylene -diamine	4.0	4.0	antiozonant
Microcrystalline wax	1.0	1.0	processing aid and finish
Mixed process oil	5.0	7.0	softener
HAF	50	—	reinforcing filler
ISAF	—	65	reinforcing filler
S	2.5	1.8	vulcanizing agent
Substituted benzothiazole-2-sulfenamide	0.5	1.5	accelerator
N-nitrosodiphenylamine (NA)	0.5	—	retarder
Total weight	173.5	220.4	
Specific gravity	1.12	1.13	

2. Put the following expressions into Chinese

    rubber formulation   natural rubber   synthetic rubber

    rubber blend   vulcanizing system   vulcanizing agent

    carbon black   calcium carbonate   resistance to fatigue cracking

    blowing agent   cellular rubber   fire retardant

3. Translate the following into Chinese

    The additives in a rubber compound may vary from 2%～3% (in the case of a rubber band) to over 60% by weight and will include some or all of the following.

    Curatives   Active chemicals which bring about the cross-linking of the long chain rubber polymer. Sulphur was the first to be discovered and is still commonly used.

Accelerators Chemicals which vary the speed and timing of the curing reaction.

Reinforcing Fillers Materials which increase the strength of the material. Carbon black and silicas are the most commonly used.

Fillers Relatively inert chemicals, such as clays, which increase the bulk of the compound.

Pigments Added to produce specified colours. They can only be used with compounds which do not contain carbon black.

Plasticisers Added to aid processability or to produce specified properties.

Anti-Oxidants Anti-Ozonants Chemicals which are added to help the compound resist surface attack, especially by ozone.

Process Aids Resins, soaps, low-weight polyethylene.

[Reading Material]

Importance of Rubber

Civilization as we know it today is wholly dependent upon rubber. It is a material of myriad uses, totally unlike anything the world had previously known. It enters in a thousand ways into the fabric of our daily lives. It is indispensable in transportation, in communication, in furnishing us with light and power, in cushioning our bodies and protecting our senses from the jars and jolts, the noise and tumult of modern life. Foe of corrosion, abrasion and vibration it aids industry in avoiding the payment of hundreds of millions of dollars which these looters annually attempt to exact. Even in helping us to spend our leisure, rubber is essential, for there are few action games in which a rubber ball has no part. It is a servant that follows us, literally, from the cradle to the grave. We are ushered into the world by the rubber-covered hands of a doctor in surroundings made sterile and quiet by this substance, and we make our exit in a rubber-gasketed coffin hauled by a rubber-tired hearse.

– The late Ralph Wolf, chemist and author, in an article in the October 1964 edition of "Rubber World."

Words and Expressions

civilization	[ˌsivilaiˈzeieiʃən;-liːz-]	n.	文明，文化，文明社会
wholly	[ˈhəuli]	adv.	整个，统统，全部
dependent	[diˈpendənt]	adj.	依赖的，由…决定的
material	[məˈtiəriəl]	n.	材料，原料，物资
myriad	[ˈmiriəd]	adj.	无数的，种种的
previously	[ˈpriːvjuːsli]	adv.	先前，以前
fabric	[ˈfæbrik]	n.	构造
indispensable	[ˌindisˈpensəbl]	adj.	不可缺少的，绝对必要的
transportation	[ˌtrænspɔːˈteiʃən]	n.	运输，运送
communication	[kəˌmjuːniˈkeiʃn]	n.	传达，信息，交通，通信
furnish	[ˈfəːniʃ]	vt.	供应，提供，装备
		v.	供给
cushioning			减震，缓冲

jar	[dʒɑː]	n.	震动，刺耳声，争吵
jolt	[dʒəult]	v. / n.	摇晃
tumult	[ˈtjuːmʌlt]	n.	吵闹，骚动，拥挤，混乱
vibration	[vaiˈbreiʃən]	n.	振动，颤动，摇动，摆动
looter		v.	掠夺，抢劫，强夺
literally	[ˈlitərəli]	adv.	逐字地
cradle	[ˈkreidl]	n.	摇篮
grave	[greiv]	n.	墓穴，坟墓
usher	[ˈʌʃə]	vt.	引导，展示
sterile	[ˈsterail]	adj.	消过毒的
coffin	[ˈkɔfin]	n.	棺材
haul	[hɔːl]	vt.	拖拉，拖运
hearse	[həːs]	n.	柩车，灵车

Unit 2 Rubbers

Functions

Rubber is a high molecular mass polymer, or macromolecule, that provides the basic elastomeric features of the end product. It is responsible for the high extensibility and the rapid recovery of shape after release from deformation. Molecular masses are typically in excess of 100000 and can reach several million the case of raw natural rubber. The long polymer chains are flexible and coiled, allowing extension by small deforming forces.

Types

Vulcanized Rubbers

There are three basic categories of vulcanizable (that is, crosslinkable) rubbers.

General purpose rubbers

[1] These are the most widely used rubbers, largely because of their uptake in pneumatic tyres, and account for about 80% of total rubber consumption. They comprise the following:

• natural rubber (chemically *cis*-1,4-polyisoprene and produced mainly from the botanical source, Hevea brasiliensis)

• butadiene rubber

• styrene-butadiene rubber (SBR, by far the most widely consumed synthetic rubber)

• isoprene rubber (the synthetic rubber equivalent of natural rubber)

General purpose rubbers have main chain unsaturation, are usually sulphur vulcanized, and have no oil resistance.

Special purpose rubbers

These rubbers have special attributes that favour their use in a range of applications. They account for up to 15% of total rubber consumption. The most commonly used types are:

Ethylene-propylene polymers (EPM and EPDM)	mainly for ozone and heat resistance
Acrylonitrile-butadiene copolymers (nitrile rubbers)	mainly for oil resistance
Isobutylene-isoprene opolymers (butyl rubbers)	mainly for high damping and low gas permeability
Chloroprene rubbers	mainly for weathering resistance
Acrylic rubbers	mainly for oil and heat resistance
Polyurethanes	mainly for foam production and casting

[2] EPDM is a terpolymer of ethylene, propylene and a diene monomer that introduces side chain unsaturation suitable for sulphur vulcanization. It is becoming more and more a general-purpose rubber because of its uptake in the automotive industry and other sectors.

三元共聚物/二烯烃
侧链
汽车工业

Speciality rubbers

特种胶

These are normally small volume and relatively expensive materials having exceptional special properties and account for no more than around 5% of total rubber consumption. They include:

fluorocarbon rubbers	oil and high temperature resistance
silicone rubbers	low temperature and high temperature resistance; chemical inertness
polysulphide rubbers	high oil resistance

氟橡胶/耐高温
硅橡胶
化学惰性
聚硫橡胶

New Words

elastomeric	[i,læstə'merik]	adj.	弹性体的
extensibility	[iks,tensə'biliti]	n.	伸长(性)
extension	[iks'tenʃən]	n.	延长，伸长
recovery	[ri'kʌvəri]	n.	恢复
release	[ri'li:s]	vt.	释放
deformation	[,di:fɔ:'meiʃən]	n.	变形
category	['kætigəri]	n.	种类
pneumatic	[nju(:)'mætik]	adj.	充气的
unsaturation	['ʌnsætʃə'reiʃən]	n.	不饱和
consumption	[kən'sʌmpʃən]	n.	消费，消费量
cis-			顺式
isoprene	['aisəupri:n]n.	n.	异戊二烯
polyisoprene	[pɔli'aisəupri:n]	n.	聚异戊二烯
Hevea brasiliensis			巴西三叶胶树
butadiene	[,bju:tə'daii:n]	n.	丁二烯
styrene	['stairi:n]	n.	苯乙烯
ethylene	['eθili:n]	n.	乙烯，亚乙基
nitrile rubber(NBR)	['naitrail]	n.	丁腈胶
isobutylene	[,aisəu'bju:tili:n]	n.	异丁烯
butyl	['bju:til]	n.	丁基
damping	['dæmpiŋ]		阻尼，衰减
gas permeability	[,pə:miə'biliti]		透气性
chloroprene	['klɔ:rəpri:n]	n.	氯丁二烯
acrylic rubber (ACM)	[ə'krilik]		丙烯酸酯橡胶
foam	[fəum]	n.	泡沫橡胶，泡沫塑料
casting	['kɑ:stiŋ]	n.	浇铸

terpolymer	[təˈpɔlimə]	n.	三元共聚物
diene	[ˈdaiiːn]	n.	二烯(烃)
monomer	[ˈmɔnəmə]	n.	单体
fluorocarbon	[ˌfluː(ː)ərəˈkɑːbən]	n.	碳氟化合物
silicone	[ˈsilikəun]	n.	聚硅氧烷，硅酮(旧称)
polysulphide	[ˌpɔliˈsʌlfaid, -fid]	n.	多硫化物

Notes

[1] These are the most widely used rubbers, largely because of their uptake in pneumatic tyres, and account for about 80% of total rubber consumption. 通用胶用量大，主要用在充气轮胎中，大约占总橡胶消耗量的 80%。

[2] EPDM is a terpolymer of ethylene, propylene and a diene monomer that introduces side chain unsaturation suitable for sulphur vulcanization. 三元乙丙胶是单体乙烯、丙烯和二烯烃的三元共聚物，二烯烃能在分子侧链中引入不饱和结构以利于采用硫黄硫化。

Exercises

1. Translate the following into Chinese

Elastomers

A brief outline of the main synthetic rubbers is presented here. We will refer to the various elastomers by their commonly-used names. In general, the compounding of synthetics follows the same pattern as for natural rubber—differences will be pointed out as we come to them.

SBR　This is a general purpose rubber mainly used in the treads and soles. The abbreviations LTP and OEP both refer to SB rubbers; LTP is a low temperature polymer or 'cold' rubber since it is made by a reaction carried out at a low temperature(41℉). OEP denotes oil extended polymer since there is an addition of oil during the manufacture of the rubber.

Butyl　This was originally a special purpose rubber noted for its very low permeability to air and other gases. It finds its greatest use in pneumatic tyre inner tubes and similar applications. It has excellent resistance to attack by ozone, and is being developed for many uses in which resistance to weathering is required.

Nitrile rubber　This rubber has extremely high resistance to attack by oil and is therefore used in making oil seals , gaskets , and similar products.

Neoprene　This rubber also has good oil resistance combined with excellent resistance to weathering.

Silicone rubber　The outstanding feature of this synthetic is its resistance to extremes of temperature. It remains flexible at temperatures as low as −80℉ and can withstand heating for long periods at temperature up to 500℉ without great loss of physical properties.

2. Describe the following abbreviation into English

　　NR　　SBR　　IR　　BR　　CR　　IIR　　NBR　　EPM　　EPDM　　ACM　　Q

3. Put the following words or expressions into Chinese

molecular mass	macrmolecule	extensibility
recovery	raw natural rubber	pneumatic tire
cis-1,4-polyisoprene	main chain	sulfur vulcanization
oil resistance	ozone resistace	gas permeability

[Reading Material]

Other Types of Rubbers

Thermoplastic rubbers

One further category comprises thermoplastic rubbers or elastomers which depend for their elasticity not on crosslinks but on zones of hard or crystalline polymers. These zones soften or melt at elevated temperatures allowing processing and shaping of products, but reform at ambient temperatures to inhibit flow and restore elasticity.

Thermoplastic rubbers, or TPEs, are prepared by two routes. The first is direct synthesis by block copolymerization or other means. The other is by the blending or alloying of a suitable rubber and plastics material.

Recycled rubbers

The industry also uses recycled rubber from factory scrap and end-of-life products. There are two basic forms-ground rubber waste in which the vulcanized state is retained and reclaimed rubber in which substantial devulcanization has taken place. Much effort has been made in recent years to increase quality so that a higher proportion of recycled rubber can be used in compounds without undue sacrifice in properties. By definition thermoplastic rubbers can be recycled without special treatment.

Blends

In many applications blends of different rubber types are used in order to take some advantage of each component. For example EPDM may be blended with SBR to confer some weathering resistance.

Liquid form

Some vulcanizable rubbers, notably natural, chloroprene, nitrile and SBR, are also available in latex form and are used for such operations as dipping, casting and foaming. Some polyurethane rubbers are available in liquid form suitable for casting.

Words and Expressions

thermoplastic rubber (TPR)			热塑性橡胶
thermoplastic elastomer (TPE)			热塑性弹性体
elasticity	[ilæs'tisiti]	*n.*	弹性
elevated temperature			高温
ambient temperature			环境温度
block	[blɔk]	*n.*	嵌段
copolymerization	[kəuˌpɔlimərai'zeiʃən]	*n.*	共聚合(作用)

alloying	[ˈæloiiŋ]	n.	合金
recycle	[ˈriːˈsaikl]	v.	反复应用
scrap	[skræp]	n.	小片，废料
retain	[riˈtein]	vt.	保持，保留
reclaimed rubber (RR)			再生胶
substantial	[səbˈstænʃəl]	adj.	实质的，真实的
devulcanization	[diːvʌlkənaiˈzeiʃən]	n.	脱硫(作用)
undue	[ˈʌnˈdjuː]	adj.	不适当的
sacrifice	[ˈsækrifais]	n.	牺牲，献身
		v.	牺牲
definition	[ˌdefiˈniʃən]	n.	定义，解说
treatment	[ˈtriːtmənt]	n.	加工，处理
component	[kəmˈpəunənt]	n.	成分
dip	[dip]	v.	浸渍

Unit 3 Vulcanizing System

Functions

[1] The vulcanizing system is the combination of vulcanizing agent with activators, accelerators or coagents needed to convert the rubber compound from an essentially plastic state to a shaped elastomeric thermoset. The system is added to the rubber polymer during mixing and remains inactive during subsequent processing and fabrication operations. Vulcanization then takes place, usually by the application of heat to activate the system. Crosslinks are inserted between adjacent polymer chains to form a three-dimensional network which prevents further flow.

Stiffness and resilience (elasticity) increase with crosslink density, whereas set, stress-relaxation and creep all decrease. For most rubbers, tensile strength reaches a peak and then falls as crosslink density is raised.

Types

[2] Choice of vulcanizing or crosslinking agent depends on the rubber type and to a lesser extent on the combination of properties required for the product, especially dynamic requirements, fatigue, heat and reversion resistance.

Sulphur

Elemental sulphur is the predominant vulcanizing agent for general-purpose rubbers. It is used in combination with one or more accelerators and an activator system comprising zinc oxide and a fatty acid (normally stearic acid). The most popular accelerators are delayed-action sulphenamides, thiazoles, thiuram sulphides, dithocarbamates and guanidines. Part or all of the sulphur may be replaced by a sulphur donor such as a thiuram disulphide.

[3] The accelerator determines the rate of vulcanization, whereas the accelerator to sulphur ratio dictates the efficiency of vulcanization and, in turn, the thermal stability of the resulting vulcanizate.

In natural rubber an accelerator to sulphur ratio typically of 1:5 is called a conventional vulcanizing system and it gives a network in which about 20 sulphur atoms are combined with the rubber for each inserted chemical crosslink. [4] Most of the crosslinks are polysulphidic (ie with a bridge of not less than three sulphur atoms) and a high proportion of the sulphur is in the form of cyclic sulphide main chain

modifications. This combination provides good <u>mechanical properties</u> and excellent <u>low temperature resistance</u>, but <u>polysulphidic crosslinks</u> are thermally unstable and reversion can occur at high vulcanizing temperatures and high service temperatures. 　力学性能
　耐寒性/多硫键

An accelerator to sulphur ratio of 5:1 is typical of an <u>efficient vulcanizing</u> (EV) system where no more than 4~5 sulphur atoms are combined with the rubber for each chemical crosslink. [5]Most of the crosslinks at <u>optimum cure</u> are monosulphidic or disulphidic and only a relatively small proportion of the sulphur is wasted in main chain modifications. This combination provides very much enhanced thermal stability, both under aerobic and <u>anaerobic conditions</u>, but some mechanical properties may be impaired. 　有效硫黄硫化体系

　正硫化

　无氧条件

An intermediate accelerator to sulphur ratio of 1:1 is typical of a <u>semi-efficient vulcanizing</u> (semi-EV) system and provides properties between those of conventional and EV systems. 　半有效硫黄硫化体系

The same principles apply to synthetic rubbers, although the optimum accelerator to sulphur ratio may not be the same as in natural rubber.

New Words

coagent	[kəu'eidʒnt]	n.	活性助剂
thermoset	['θə:məset]	n.	热固性树脂, 热固性塑料
		adj.	热固性的
inactive	[in'æktiv]	adj.	无活性的
fabrication	[ˌfæbri'keiʃən]	n.	制作, 装配
activate	['æktiveit]	vt.	活化
insert	[in'sə:t]	vt.	插入, 嵌入
adjacent	[ə'dʒeisnt]	adj.	邻近的, 接近的
network	['netwə:k]	n.	交联网
resilience	[ri'ziliəns]	n.	回弹性
set	[set]	n.	永久变形
creep	[kri:p]	n.	蠕变
predominant	[pri'dɔminənt]	adj.	主要的
zinc oxide			氧化锌（ZnO）
fatty acid			脂肪酸
stearic acid			硬脂酸(SA)
delayed-action			迟效性
sulphenamide			次磺酰胺
thiazole	['θaiəˌzəul]	n.	噻唑
thiuram	['θijuræm]	n.	秋兰姆
sulphide	['sʌlfaid]	n.	硫化物

dithocarbamate			二硫代氨基甲酸盐
guanidine	[ˈgwænədiːn]	n.	胍
conventional	[kənˈvenʃənl]	adj.	常规的，传统的
polysulphidic			多硫的
not less than		n.	至少
cyclic	[ˈsaiklik]		环状的
modification	[ˌmɔdifiˈkeiʃən]	n.	改性
reversion	[riˈvəːʃən]	n.	返原，回复，复原
efficient	[iˈfiʃnt]	adj.	有效率的
monosulphidic			单硫的
disulphidic			二硫的
proportion	[prəˈpɔːʃən]	n.	比例，部分
aerobic	[ˌeiəˈrəubik]	adj.	有氧气的
intermediate	[ˌintəˈmiːdjət]	adj.	中间的
semi-	[ˈsemi]		半
optimum	[ˈɔptiməm]	adj.	最适宜的

Notes

[1] The vulcanizing system is the combination of vulcanizing agent with activators, accelerators or coagents needed to convert the rubber compound from an essentially plastic state to a shaped elastomeric thermoset. 硫化体系是由硫化剂、促进剂、活性剂或活性助剂组成的体系，它们能将橡胶由塑性状态转变为固定形状的热固性弹性体。

[2] Choice of vulcanizing or crosslinking agent depends on the rubber type and to a lesser extent on the combination of properties required for the product, especially dynamic requirements, fatigue, heat and reversion resistance. 硫化剂的选择主要取决于胶种，另外也和制品所要求的各项性能有关，特别是动态性能、抗疲劳性能、耐热性和抗硫化返原性。

[3] The accelerator determines the rate of vulcanization, whereas the accelerator to sulphur ratio dictates the efficiency of vulcanization and, in turn, the thermal stability of the resulting vulcanizate. 促进剂的量决定了硫化快慢，而促进剂与硫黄的比率决定了硫化效率，进而又影响硫化胶的耐热性。

[4] In natural rubber an accelerator to sulphur ratio typically of 1:5 is called a conventional vulcanizing system and it gives a network in which about 20 sulphur atoms are combined with the rubber for each inserted chemical crosslink. Most of the crosslinks are polysulphidic (ie with a bridge of not less than three sulphur atoms) and a high proportion of the sulphur is in the form of cyclic sulphide main chain modifications. 对天然胶来说，促进剂与硫黄的比例为1:5 称为普通硫黄硫化体系，所得到的交联网中每个化学交联键大约结合有 20 个硫原子。多数交联键都是多硫键（即交联键间不少于 3 个硫原子），硫黄大部分在分子主链中以硫环的形式存在。

[5] Most of the crosslinks at optimum cure are monosulphidic or disulphidic and only a relatively small proportion of the sulphur is wasted in main chain modifications. 达到正硫化时大部分交联键都是单硫键或双硫键，只有少部分硫黄浪费在分子主链中(形成硫环)。

Exercises

1. Fill in the blanks with proper words or expressions

   Vulcanizing system is the combination of _____, _____, _____ and/or retarders, which can make the rubber become _____ through the process of _____ . By far the most important vulcanizing agent is_____, and the most popular accelertors are _____, _____, _____, _____etc., and the most commonly used activators are _____and_____. Accelerators added to the rubber can _____the rate of vulcanization, _____ the temperature of vulcanization, _____ the amount of sulfur and_____ the physical properties of vulcanizates. According to the ratio of sulfur to accelerator, the sulfur vulcanization system can be classified as _____, _____and_____.

2. Put the following words or expressions into English

交联	回弹性	弹性	永久变形	蠕变	应力松弛
拉伸强度	交联密度	氧化锌	硬脂酸	脂肪酸	硫化秋兰姆
单硫键	双硫键	多硫键	普通硫化体系	有效硫化体系	半有效硫化体系

3. Translation

   In 1839 Charles Goodyear invented a process called *vulcanization*, which made rubber more durable and more useful at hot and cold temperatures. Vulcanization crosslinks rubber. That is, it ties all the chainlike rubber molecules together to form one big molecule.

   This does a lot for the rubber, but it can also make rubber difficult to process. Think about it. A material made of one molecule can't really flow into a mold, and it can't be shaped or worked very easily. A material made of crosslinked rubber has to be shaped before it is crosslinked. Crosslinking also makes rubber difficult to recycle.

   Some kinds of synthetic rubber aren't crosslinked, however. They can be molded and remolded again and again. They are called *thermoplastic elastomers*. *Thermo-* means "heat", and *plastic* means "moldable". *Elastomer* is just a fancy word that means "rubber." So a thermoplastic elastomer is a rubber that can be molded when it is heated.

[Reading Material]

Other Types of Vulcanizing Agents

Peroxides

Organic peroxides are used (a) to crosslink rubbers having no main chain unsaturation and (b) to crosslink unsaturated rubbers when sulphur vulcanization is found wanting. A peroxide initiates crosslinking through a free radical reaction. In essence a thermally liberated peroxy radical abstracts a labile hydrogen from the polymer chain and then crosslinking is effected by the combination of two adjacent polymer free radicals. In practice the reaction is more complicated because coagents are often used with the peroxide to enable a reduction in peroxide levels and improve processing safety.

A peroxide is characterized by a half life at any given temperature and it is advised that the rubber is vulcanized for the equivalent of at least five half lives to ensure only a trace of unreacted peroxide remains. Residual peroxide can initiate oxidation as well as introducing unwanted additional crosslinks in product service.

The principal classes of peroxide crosslinking agents are dialkyl and diaryl peroxides, peroxyketals and peroxyesters.

Peroxide vulcanization has only a few applications in natural rubber, SBR and other general purpose rubbers because the mechanical and dynamic properties are generally not as good as in sulphur vulcanizates.

Other vulcanizing agents

These ingredients include amines for the crosslinking of fluorocarbon rubbers, metal oxides for chlorine-containing rubbers (notably zinc oxide for chloroprene rubber) and phenol-formaldehyde resins for the production of heat-resistant butyl rubber vulcanizates.

Some rubbers, such as natural rubber, can also be crosslinked by selective high energy radiation. This process has the advantage that additives can be kept to a minimum level-which is important for some regulated medical and surgical applications. Even so, an antioxidant is often necessary to guard against oxidation.

Words and Expressions

peroxide	[pəˈrɔksaid]	n.	过氧化物
organic	[ɔːˈgænik]	adj.	有机的
wanting	[ˈwɔntiŋ]	adj.	欠缺的, 没有的
initiate	[iˈniʃieit]	vt.	引发
peroxy radical			过氧自由基
labile	[ˈleibail]	adj.	不安定的
complicated	[ˈkɔmplikeitid]	adj.	复杂的, 难解的
half life			半衰期
equivalent	[iˈkwivələnt]	adj.	相等的, 相当的
trace	[treis]	n.	微量
remains	[riˈmeins]	n.	残余
residual	[riˈzidjuəl]	adj.	剩余的, 残留的
principal	[ˈprinsəp(ə)l, -sip-]	adj.	主要的, 首要的
dialkyl	[daiˈælkil]		二烷基
diaryl			二芳基
peroxyketal			过氧化酮
peroxyester			过氧化酯
metal oxide			金属氧化物
chlorine	[ˈklɔːriːn]	n.	氯
regulate	[ˈregjuleit]	vt.	管制, 控制
medical	[ˈmedikəl]	adj.	内科的, 医学的
surgical	[ˈsəːdʒikəl]	adj.	手术上的, 外科的

Unit 4 Antidegradants

Function

Antidegradants, alternatively known as protective agents or antiagers, inhibit rubber degradation, primarily in product service but also sometimes in the raw rubber state.

The most commonly used type is an antioxidant, which is added to retard oxidation. It does so in one of two ways. [1]The first is to decompose potentially damaging hydroperoxides into harmless reaction products no longer able to initiate oxidation; the second is to interfere with the propagation process by removing peroxy free radicals and therefore break the chain reaction. Some antioxidants are also able to inhibit oxidation initiated by UV radiation (photo-oxidation) and by traces of redox heavy metals such as copper, iron and manganese (metal ion catalysis).

Types

Antioxidants

These fall into three categories are as follows.

Amine derivatives

These include substituted p-phenylenediamines, substituted diphenylamines and ketone-amine condensates. They are usually highly effective inhibitors of thermal oxidation and some of the p-phenylenediamines are powerful anti-flex cracking agents. Several are also effective against metal ion catalysis. However, amines undergo discolouration, cause staining of light-coloured surfaces and can occasionally promote photo-oxidation. They are therefore largely confined to black-filled rubbers.

Phenolic derivatives

These comprise substituted (hindered) phenols, and bisphenols in the form of phenyl alkanes, phenolic sulphides and polyphenols. They are essentially non-staining and non-discolouring and so are commonly used in non-black rubbers. They are not effective against ozone attack and have only a small anti-flex cracking activity.

Antiozonants

Function

Chemical antiozonants are added to unsaturated rubbers to prevent surface cracking caused by atmospheric ozone attack of the double bond. [2]The mechanism is very complex and still not fully understood but key features are the ability of antiozonants to diffuse to the rubber surface to form a <u>protective layer</u> and to reduce the rate of <u>crack growth</u>.

保护层
裂纹增长

Types

Substituted p-phenylenedamines are the most powerful <u>chemical antiozonants</u>, but because of staining and discolouration their use are again confined essentially to black filled rubbers. In general-purpose rubbers, <u>dialkyl</u> substituted *p*-phenylenediamines are capable of enhancing the threshold <u>tensile strain</u> necessary to initiate cracks. Alkyl-phenyl *p*-phenylenediamines are effective in reducing the rate of crack growth, an important feature in dynamically deformed rubber.

化学抗臭氧剂

二烷基
拉伸应变

<u>Hydrocarbon waxes</u> are used as <u>physical antiozonants</u> through their ability to diffuse to the rubber surface to form an <u>ozone-impermeable layer</u>. They are only effective under <u>static</u> strain conditions but are often used in combination with a chemical antiozonants for <u>all-round protection</u>.

烃蜡/物理抗臭氧剂
臭氧保护层

全面防护

New Words

protective	[prəˈtektiv]	adj.	保护的
antiager			防老剂
primarily	[ˈpraimərili]	adv.	首先, 起初, 主要地
raw	[rɔ:]	adj.	未加工的, 生疏的
decompose	[ˌdi:kəmˈpəuz]	v.	分解, (使)腐烂
hydroperoxide	[ˌhaidrəpəˈrɔksaid]	n.	氢过氧化物, 过氧化氢物
interfere	[ˌintəˈfiə]	vi.	干涉, 干预, 妨碍
removing	[riˈmu:viŋ]		消除, 除去
chain reaction			链式反应, 连锁反应
UV(ultra violet)		adj.	紫外的
radiation	[ˌreidiˈeiʃən]	n.	辐射, 放射
redox	[ˈredɔks]	n.	氧化还原作用
manganese	[ˌmæŋgəˈni:z, ˈmæŋgəni:z]	n.	锰(Mn)
ion	[ˈaiən]	n.	离子
catalysis	[kəˈtælisis]	n.	催化作用
derivative	[diˈrivətiv]	n.	衍生物
p-phenylenediamine	[ˈfenli:nˈdaiəmin]		对苯二胺
diphenylamine	[daiˌfenləˈmi:n]	n.	二苯胺
ketone	[ˈki:təun]	n.	酮
condensate	[kɔnˈdenseit]	n.	缩合物
inhibitor	[inˈhibitə(r)]	n.	抑制剂

staining	['steiniŋ]		污染，着色
confine oneself to			只涉及，只限于
hinder	['hində]	v.	阻碍，打扰
phenol	['fi:nəl]	n.	苯酚
bisphenol			双酚
phenyl	['fenəl, 'fi:nəl]	n.	苯基
alkane	['ælkein]	n.	烷烃
polyphenol	[,pɔli'fi:nɔl]		多元酚
mechanism	['mekənizəm]	n.	机理，机制
layer	['leiə]	n.	层
threshold	['θreʃhəuld]	n.	开始，开端
hydrocarbon	['haidrəu'kɑ:bən]	n.	烃，碳氢化合物

Notes

[1] It does so in one of two ways. The first is to decompose potentially damaging hydroperoxides into harmless reaction products no longer able to initiate oxidation; the second is to interfere with the propagation process by removing peroxy free radicals and therefore break the chain reaction. 防老剂的作用方式有两种：一是将有害的氢过氧化物分解成无害的不会引起氧化反应的产物；另一个是通过消除过氧自由基以干预自由基的继续反应，从而使连锁反应终止。

[2] The mechanism is very complex and still not fully understood but key features are the ability of antiozonants to diffuse to the rubber surface to form a protective layer and to reduce the rate of crack growth. （抗臭氧剂的）作用机理非常复杂还不完全清楚，但抗臭氧剂的主要特征表现为能扩散到橡胶表面，形成保护层从而降低裂纹的增长速率。

Exercises

1. Translate the following into Chinese

Improvement of Ozone Resistance of NBR

Although improvement in the ozone resistance of nitrile rubber by chemical antioxidants has not been advanced since 1963, there had been considerable activity on blending NBR with saturated polymers, such as PVC and EPDM, as a physical means of improving ozone resistance without discoloring the polymer.

One might expect any saturated rubbery polymer to improve ozone resistance, but Khanin and coworkers claim that only EPDM imparts ozone resistance in 30:70 blends with SBR or NBR. This is attributed to the ability of EPDM to migrate to the polymer surface. Khanin's work might bear repeating since PVC is certainly effective at 30% concentration.

2. Put the following expressions into Chinese

protective agent	peroxy free radical	chain reaction	UV radiation
heavy metal	ketone-amine	black-filled rubber	substituted pheol
double bond	non-black rubber	unsaturated rubber	protective layer
static strain	hydrocarbon wax	non-staining	non-discolouring

3. Match the columm A with B

Columm A	Columm B
oxidation	臭氧老化
ozone attack	抗降解剂
photo-oxidation	抗氧化剂
thermal oxidation	抗臭氧剂
antioxidant	抗屈挠龟裂剂
antiozonant	氧化
anit-flex cracking agent	光氧化
antiager	金属离子催化老化
anti-flex cracking	屈挠龟裂
metal ion catalysis ageing	热氧化
antidegradant	胺类防老剂
amine antioxidants	防老剂
mixture of antioxidants	酚类防老剂
phenol antioxidants	物理防老剂
chemical antioxidant	化学防老剂
physical antioxidant	混合防老剂

[Reading Material]

The Ageing of Rubber Vulcanizates

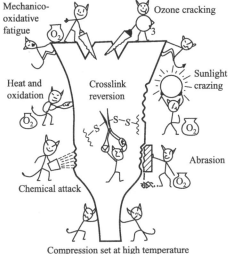

Types of degration in natural and synthetic rubber vulcanizates

The factors which influence the life of an elastomeric product fall into two basic categories which can be called "product characteristics" and "ageing processes".

Once the product exists, this illustration shows what can happen to it and again the potential for damage falls into several groups. Abrasion and external chemical attack may be unavoidable during service life but should not occur on storage. Temperature, sunlight and stresses can all accelerate or initiate the chemical changes which are waiting to happen.

These changes, collectively known as 'ageing' involve three distinct and potentially co-synchronous mechanistic routes for most (sulphur) vulcanizates. These can be identified as: continuing sulphur chemistry, shelf ageing, atmospheric ageing.

Words and Expressions

category	[ˈkætigəri]	n.	种类，类别
illustration	[ˌiləsˈtreiʃən]	n.	图表，插图
potential	[pəˈtenʃ(ə)l]	adj.	潜在的，可能的
damage	[ˈdæmidʒ]	n.	损害，伤害
external	[eksˈtəːnl]	adj.	外部的，表面的
		n.	外部，外面
chemical attack			化学腐蚀
unavoidable	[ˌʌnəˈvɔidəbl]	adj.	不可避免的
collectively		adv.	全体地，共同地
distinct	[disˈtiŋkt]	adj.	明显的，独特的
synchronous	[ˈsiŋkrənəs]	adj.	同时的，同步的
mechanistic	[ˌmekəˈnistik]	adj.	机械论的
route	[ruːt]	n.	路线，路程
shelf ageing			储存老化

Unit 5　Fillers

Functions

Fillers are added to rubber for a variety of purposes, of which the most important are <u>reinforcement</u>, <u>increase in stiffness</u>, <u>reduction in material costs</u> and <u>improvements in processing</u>.

[1] Reinforcement is primarily the enhancement of strength and strength-related properties, including <u>resistance to abrasive wear</u>, and it is especially important for rubbers that are inherently weak through their inability to <u>strain-crystallize</u>. Reinforcement is accompanied by stiffening or increase in <u>modulus</u>. [2]A <u>gum</u> (ie. unfilled) vulcanizate for most types of rubber has a hardness normally no higher than 40 <u>Shore A</u> or <u>IRHD</u> units, equivalent to a <u>Young's modulus</u> of about 1.5MPa, whereas a <u>tyre tread</u> has a hardness in the order of 60~70units and a <u>rubber sole</u> has a hardness around 80units. Many products such as <u>mountings</u>, <u>gaskets</u> and <u>seals</u> are available in a wide range of hardness levels.

Many fillers aid rubber processing by reducing <u>elastic recovery</u> and by <u>improving extrudability</u>, sheeting and other operations. Other fillers confer special attributes to the rubber such as <u>electrical conductivity</u>, <u>chemical resistance</u> and <u>flame retardance</u>.

Most fillers are <u>particulate</u> in nature, with reinforcing types having <u>particle sizes</u> as small as 20nm. <u>Fibrous fillers</u> are also available with the advantage that stiffness can be enhanced in one direction relative to the other. <u>Reinforcing resins</u> (eg based on <u>phenol-formaldehyde</u>) are also used for relatively hard vulcanizates (ie. above 80 shore A or IRHD).

Types

There are basically two categories of particulate fillers - carbon black and <u>mineral or 'white' fillers</u>. They are often used in combination.

Carbon black

This is the most widely used rubber filler, and a wide range of grades is available. The filler consists of aggregates of approximately spherical <u>primary particles</u> and these two characteristics-aggregate 'structure' and particle size-determine the effect the carbon black has in rubber.

Grades are now classified by the 4-digit nomenclature system standardized in ASTM D. 1765. [3]The first digit is a letter (usually N) denoting the effect on the rate of vulcanization, the second digit (the first of three numbers) is related to particle size and the other digits are arbitrary. Reinforcing grades are in the N100 to N300 series, semi-reinforcing grades in the N500 to N700 series and diluent 'thermal' grades are designated N880 and N990. [4]The reinforcing grades are used in tyre components, conveyer belting and other products requiring tear and abrasion resistance, whilst the semi-reinforcing grades are usually employed in engineering components and seals. Most blacks are now produced by the furnace process and thus known as furnace blacks.

Mineral fillers

The principal types are as follows:

Silicas

Fine particles size types can be as reinforcing as carbon blacks. Blends with carbon black are being increasingly used in tyre treads to enhance resistance to wet grip and to reduce rolling resistance (with a consequent saving in energy consumption).

Silicates

Calcium, magnesium and aluminium silicates are available. Aluminum silicate offers antistatic properties as an alternative to reinforcing carbon blacks.

Clays

Hard and soft clays are used in a wide variety of general-rubber goods.

Calcium carbonates

Both chemically precipitated and ground naturally occurring types are used, with applications ranging from hot water bottles to flooring. Particles sizes of diluent calcium carbonate fillers or whiting can be as high as 100μm.

Other types of filler

[5] Magnesium carbonate is suitable for translucent rubbers as an alternative to silica, barium sulphate is used for its high chemical resistance, hydride alumina is a flame retardant, and mica offers high gas impermeability because of its distinctly lamellar structure.

New Words

reinforcement	[ˌriːinˈfɔːsmənt]	n.	加强，补强
abrasive wear			磨损
inherently	[inˈhiərəntli]	adv.	天生地，固有地
accompany	[əˈkʌmpəni]	vt.	伴随

IRHD(International Rubber Hardness Degree)			国际橡胶硬度
modulus	[ˈmɔdjuləs]	n.	模量
in the order of			大约
sole	[səul]	n.	鞋大底
mounting	[ˈmauntiŋ]	n.	减震器
gasket	[ˈgæskit]	n.	垫圈，衬垫
seal	[siːl]	n.	密封圈
confer	[kənˈfəː]	vt.	授予，赠与
attribute	[əˈtribju(ː)t]	n.	属性，品质，特征
fibrous	[ˈfaibrəs]	adj.	纤维状的
mineral	[ˈminərəl]	adj.	矿质的，无机的
aggregate	[ˈægrigeit]	n.	集合体
		v.	聚集，集合
approximately	[əprɔksiˈmətli]	adv.	近似地，大约
spherical	[ˈsferikəl]	adj.	球的，球形的
characteristic	[ˌkæriktəˈristik]	adj.	特有的，典型的
		n.	特性，特征
classify	[ˈklæsifai]	vt.	分类
digit	[ˈdidʒit]	n.	阿拉伯数字
nomenclature	[nəuˈmenklətʃə]	n.	命名法，术语
standardized	[ˈstændəˌdaizd]		标准的
ASTM (American Society of Testing and Materials)			美国材料和试验协会
denotes	[diˈnəut]	vt.	指示，表示
arbitrary	[ˈɑːbitrəri]	adj.	任意的，武断的
semi-reinforcing			半补强
designate	[ˈdezigneit]	vt.	指定
grip	[grip]	vt.	紧握，紧夹，抓住
consequent	[ˈkɔnsikwənt]	adj.	随之发生的
silicate	[ˈsilikit]	n.	硅酸盐
magnesium	[mægˈniːzjəm]	n.	镁(Mg)
antistatic	[ˌæntiˈstætik]	adj.	抗静电的
clay	[klei]	n.	陶土
goods	[gudz]	n.	制品
carbonate	[ˈkɑːbəneit]	n.	碳酸盐
precipitate	[priˈsipiteit]	vt.	使沉淀
whiting	[ˈ(h)waitiŋ]	n.	碳酸钙，白垩
barium	[ˈbɛəriəm]	n.	钡
sulphate	[ˈsʌlfeit]	n.	硫酸盐
mica	[ˈmaikə]	n.	云母
lamellar	[ləˈmelə]	adj.	片状的，层状的

Notes

[1] Reinforcement is primarily the enhancement of strength and strength-related properties, including resistance to abrasive wear, and it is especially important for rubbers that are inherently. weak through their inability to strain-crystallize. 补强主要是提高强度或与强度相关的性能(包括耐磨性), 补强对于那些不能发生拉伸结晶而纯胶强度较低的胶料来说尤其重要。

[2] A gum (ie. unfilled) vulcanizate for most types of rubber has a hardness normally no higher than 40 Shore or IRHD units, equivalent to a Young's modulus of about 1.5MPa, whereas a tyre tread has a hardness in the order of 60 ~ 70 units and a rubber sole has a hardness around 80 units. 大多数橡胶中的纯胶硫化胶(即未填充胶)的邵氏A或国际橡胶硬度通常不超过40, 相当于杨氏模量大约为1.5MPa, 而胎面胶的硬度大约为60 ~ 70, 鞋底的大约为80。

[3] Grades are now classified by the 4-digit nomenclature system standardized in ASTM D1765. The first digit is a letter (usually N) denoting the effect on the rate of vulcanization, the second digit (the first of three numbers) is related to particle size and the other digits are arbitrary. 炭黑是按ASTM D1765标准中的四字符命名系统分类。第一个字符是字母(通常是字母N), 它表明的是炭黑对硫化速率的影响, 第二个字符(三个数字中的第一个)与炭黑的粒子大小有关, 另两字符则是任意的。

[4] The reinforcing grades are used in tyre components, conveyer belting and other products requiring tear and abrasion resistance, whilst the semi-reinforcing grades are usually employed in engineering components and seals. 补强性炭黑用在轮胎, 运输带和其他要求抗撕裂、耐磨的橡胶制品中, 而半补强炭黑则通常用作工程配件和密封件。

[5] Magnesium carbonate is suitable for translucent rubbers as an alternative to silica, barium sulphate is used for its high chemical resistance, hydride alumina is a flame retardant, and mica offers high gas impermeability because of its distinctly lamellar structure. 碳酸镁可代替白炭黑用在半透明制品中, 硫酸钡能耐化学腐蚀, 氢氧化铝能阻燃, 而云母独特的层状结构使得胶料气密性好。

Exercises

1. Translation

Reinforcement of Elastomers

When the automobile first became popular, the need to toughen tyre rubber, especially against abrasion, became obvious. Although zinc oxide had already attained widespread use as a rubber colourant, in 1905 Ditmar realized the true importance of this material as a reinforcing agent for rubber. Many industry veterans can still remember when tyres had white treads. However, such tyre treads usually lasted less than 5000 miles.

In 1904 Mote had already discovered the reinforcing value of very fine carbon blacks. Carbon black proved superior to zinc oxide for rubber reinforcement, and replaced the latter in tyres between 1910 and 1915. Tyre treads lasting up to 40000 miles are now common.

2. Describe the following symbols into English

CB   WCB $CaCO_3$ $BaSO_4$ HAF   SRF   ISAF   FT   HMF   EPC

3. Put the following words or expressions into English

填料	炭黑	白炭黑	陶土	硅酸钙
硅酸铝	硅酸镁	碳酸镁	硫酸钡	氢氧化铝
软质陶土	云母	碳酸钙	硬质陶土	硫酸钡

4. Put the following words or expressions into Chinese

resistance to abrasive wear	strain-crystallize	gum vulcanizate
Young's modulus	tyre tread	rubber sole
elastic recovery	electrical conductivity	flame retardance
reinforcing filler	phenol-formaldehyde resin	fibrous filler
mineral filler	semi-reinforcing black	thermal carbon black
furnace carbon black	antistatic properties	lamellar structure

[Reading Material]

IRHD and Shore A Scales

Q: Rubber materials and products are often classified by hardness. What is this property and what is the difference between the IRHD and Shore A scales?

A: Rubber hardness is a measure of resistance to indentation under a specified force and so is a reflection of stiffness. The International Rubber Hardness Degree (IRHD) scale is related to Young's modulus, with 0 representing a material of zero modulus and 100 one of infinite modulus. The full relationship is described in ISO 48.

By way of illustration, an elastic band will have a hardness in the range of 30 ~ 45 IRHD, a tyre tread will be in the range of 60 ~ 70 IRHD and a shoe sole in the range 70 ~ 90 IRHD. Ebonite will have a hardness above 99 IRHD.

The Shore A durometer scale is numerically similar to IRHD for most rubbers but with some materials may differ by several units. A Shore D scale is available for ebonite or hard rubber and some stiffer thermoplastic rubber grades. The Shore durometer and the IRHD version of a pocket hardness meter are described in ISO 7619.

Silica Used in Tires

Reduction of fuel consumption for cars is a major issue in the automotive industry. One of the main factors determining a car's fuel consumption is its tyres. Using silica instead of carbon black as a reinforcing filler in tyre tread compounds will lower the rolling resistance by 20% to 30%, resulting in a 5% reduction in fuel consumption for a passenger car. Using silica, however, does not come without consequences. One of the main disadvantages of silica is the mixing problem. The polar silica particles are difficult to disperse in an apolar rubber. One way to solve this mixing problem is to use coupling agents. The commonly used coupling agent is bis(triethoxysilylpropyl) tetrasulphide (TESPT).

Words and Expressions

| indentation | [ˌindenˈteiʃən] | n. | 凹痕 |
| reflection | [riˈflekʃən] | n. | 反映 |

infinite	[ˈinfinit]	adj.	无限的，无数的
illustration	[ˌiləsˈtreiʃən]	n.	说明，例证，例子
elastic band			橡皮筋
durometer	[djuəˈrɔmitə]	n.	硬度计
version	[ˈvəːʃən]	n.	版本
pocket	[ˈpɔkit]	adj.	袖珍的，小型的
issue	[ˈisjuː]	n.	论点，问题
passenger car			乘用车
come with			伴随…发生
consequence	[ˈkɔnsikwəns]	n.	结果
disadvantage	[ˌdisədˈvɑːntidʒ]	n.	不利，缺点
polar	[ˈpəulə]	adj.	极性的
apolar	[eiˈpəulə]	adj.	非极性的
coupling agent			偶联剂
bis(triethoxysilylpropyl) tetrasulphide (TESPT)			双(三乙氧基硅基丙烷)四硫化物

PART E Raw Rubbers

Unit 1 Natural Rubber

Natural rubber has been successfully used as an <u>engineering material</u> for many years. That natural rubber is the most versatile engineering material can be shown by the following properties:
- <u>hardness</u> adjustable from very soft to very hard (<u>ebonite</u>);
- appearance from <u>translucent</u> (soft) to black (hard);
- <u>electrically insulating</u> or fully <u>conductive</u>;
- <u>compounded</u> to meet almost any mechanical requirement;
- <u>silence noise</u> and <u>absorb vibration</u>;
- protect, insulate and seal;
- available in any shape and <u>surface roughness</u>.

The choice of type of natural rubber depends on the purpose for which it will be used. [1]Hence, it was felt necessary to produce rubbers that meet the requirements on the following criteria: <u>latex</u> quality, non smelly, cost competitive and good (<u>physical</u>) <u>properties</u>. The different types of NR are specified in the <u>Technical Specified Rubber (TSR)</u> scheme which was first introduced by Malaysia (<u>SMR</u>). Nowadays, also new types of NR are available such as <u>superior processing grades (SP/SA)</u>, <u>epoxidized natural rubber (ENR)</u>, <u>thermoplastic natural rubber (TPNR)</u> and <u>deproteinized natural rubber (DPNR)</u>.

[2] A feature of natural rubber is that it can be compounded to have high <u>resilience</u>, high strength and high <u>fatigue resistance</u> simultaneously. The desired physical properties can be achieved by compounding using <u>ingredients</u> such as <u>carbon black</u>, <u>softeners</u>, <u>anti-degradants</u> and a <u>vulcanization system</u>. <u>Raw natural rubber</u> is a very <u>high molecular material</u>. To mix NR and ingredients the NR has to be <u>masticated</u>. Mastication shortens rubber molecular chains, resulting in a reduced molecular weight. Only then ingredients can be homogeneously distributed into the rubber.

天然橡胶

工程材料

硬度/硬质橡胶
半透明的
电绝缘的/导电的
配合
隔音/减震

表面粗糙度

胶乳
物理性能
工艺分级橡胶
马来西亚标准胶
易操作 NR
环氧化 NR/热塑性 NR
脱蛋白 NR

回弹性/抗疲劳性

配合剂/炭黑/软化剂/防老剂/硫化体系/生胶/高分子材料/塑炼

New Words

adjustable	[ə'dʒʌstəb(ə)l]	adj.	可调整的
appearance	[ə'piərəns]	n.	外貌，外观
insulating		adj.	绝缘的
conductive	[kən'dʌktiv]	adj.	传导的
silence	['sailəns]	vt.	使沉默，使安静
		v.	压制
roughness	['rʌfnis]	n.	粗糙，粗糙程度
criteria	[krai'tiəriə]	n.	标准
smelly	['smeli]	adj.	发臭的，有臭味的
scheme	[ski:m]	n.	安排，计划，方案
SMR（Standrard Malasian Rubber）			马来西亚标准橡胶
masticate	['mæstikeit]	v.	塑炼
homogeneous	[ˌhɔməu'dʒi:njəs]	adj.	均一的，均匀的

Notes

[1] Hence, it was felt necessary to produce rubbers that meet the requirements on the following criteria: latex quality, non-smelly, cost competitive and good (physical) properties. 因而，生产出的天然胶必须符合以下标准：胶乳质量高、无味、价低和具有优异的物理性能。

[2] A feature of natural rubber is that it can be compounded to have high resilience, high strength and high fatigue resistance simultaneously. 天然胶配合后具有回弹性高、强度大和耐疲劳性好的综合性能。

Exercises

1. Translate the following into Chinese

Natural rubber is a high molecular weight hydrocarbon（烃，碳氢化合物）. It comes from the latex of many plants and is obtained by coagulation（凝聚）with chemicals, by drying and other processes. The most important rubber-yielding plant is Hevea Brasiliensis（巴西三叶橡胶树）. The rubber produced from latex contains, besides the hydrocarbon, small quantities of protein, carbohydrates（碳水化合物）, mineral salts, fatty acids etc.

Natural rubber is cis-1,4-polyisoprene with the empirical formula $(C_5H_8)_n$.

Physical properties of natural rubber may vary slightly due to the non-rubber constituents and to the crystallinity.

Natural rubber finds wide application in many areas for the following reasons: superior building tack（成型黏着性）, green stock（生胶）strength, better processing, high strength in non-black formulations, hot tear resistance, retention of strength at elevated temperatures, high resilience, low hysteresis (heat build-up), excellent dynamic properties, and general fatigue resistance.

2. Write the original words for the abbreviation

    NR   RSS   SCR   SMR   STR   ENR   TPNR

3. Put the following words or expressions into English

| 工程材料 | 高分子材料 | 天然生胶 | 热塑性天然胶 | 硬度 |
| 硬质橡胶 | 回弹性 | 耐疲劳性 | 分子量 | 塑炼 |

[Reading Material]

Natural Rubber (Polyisoprene)

The original natural material which has been in commercial use since the turn of the century. The most widely developed rubber with a huge range of compounds available. It also usually has the lowest price.

Natural rubber is an environmentally desirable material and comes from a naturally replenishable source. During its production as a tree sap (latex), it constantly absorbs carbon dioxide (a greenhouse gas) from the air. At the end of their working lives, the rubber trees are used to make furniture and are replaced with young trees for further production. Natural rubber itself is readily biodegradable and non-toxic.

Here are some of the details about Natural Rubber.

Properties
- widest range of hardnesses
- very strong (naturally self-reinforcing) and extremely resilient
- good compression set
- good resistance to inorganic chemicals

Limitations
- lack of resistance to oil and organic fluids
- relatively low maximum temperatures (75℃ continuous, 100℃ intermittent)
- poor ozone resistance, with tendency to perish in open air (can be improved to some extent by careful compounding).

Typical Applications
- components which are protected from constant air changes - i.e. inside machinery - and which do not come into contact with any oil or oil based fluids
- applications requiring strength and resistance to abrasion
- sealing and shock absorption

Words and Expressions

commercial	[kə'mə:ʃəl]	adj.	商业的, 贸易的
environmental	[in,vaiərən'mentl]	adj.	环境(产生)的, 周[包]围的
replenish	[ri'pleniʃ]	v.	补充
sap	[sæp]	n.	树液
furniture	['fə:nitʃə]	n.	家具
resilient	[ri'ziliənt]	adj.	弹回的, 有回弹力的
intermittent	[,intə(:)'mitənt]	adj.	间歇的, 断断续续的
perish	['periʃ]	vi.	腐烂, 枯萎, 毁坏

Unit 2　Butadiene Rubber

Properties and Applications

Polybutadiene (BR) is the second largest volume synthetic rubber produced, next to styrene-butadiene rubber (SBR). Consumption was about 1953000 metric tons worldwide in 1999. The major use of BR is in tires with over 70% of the polymer produced going into treads and sidewalls. Cured BR imparts excellent abrasion resistance (good tread wear), and low rolling resistance (good fuel economy) due to its low glass transition temperature (T_g). [1]The low T_g, typically < −90℃, is a result of the low "vinyl" content of BR. However, low T_g also leads to poor wet traction properties, so BR is usually blended with other elastomers like natural rubber or SBR for tread compounds. BR also has a major application as an impact modifier for polystyrene and acrylonitrile-butadiene-styrene resin (ABS) with about 25% of the total volume going into these applications. Typically about 7% BR is added to the polymerization process to make these rubber-toughened resins. Also, about 20000 metric tons worldwide of "high cis" polybutadiene is used each year in golf ball cores due to its outstanding resiliency.

High Cis Polybutadiene

The alkyllithium and transition metal catalysts make very different products. [2]The transition metal, or so called Ziegler catalysts produce very "stereoregular" BRs with one type having the main polymer chain on the same side of the carbon-carbon double bond contained in the polybutadiene backbone. This is called the cis configuration.

$$\sim\!\!\sim\!\!CH_2\underset{\underset{\displaystyle cis\text{-}1,4\text{-}}{}}{\overset{H\quad\quad H}{\underset{|\quad\quad|}{C\!=\!C}}}CH_2\!\sim\!\!\sim$$

Lithium-based Polybutadiene

The alkyllithium or "anionic" catalyst system produces a polymer with about 40% *cis*, 50% *trans* and 10% vinyl when no special polar modifiers are used in the process. Vinyl increases the T_g of the polybutadiene by creating a stiffer chain structure. Vinyl also tends to crosslink or "cure" under high heat conditions so the high vinyl polymers are less thermally stable than low vinyl. Notes below, that in vinyl units the double bonds are pendent to the main chain, giving rise to the special properties of high vinyl polymers.

$$\sim CH_2-CH\sim$$
$$|$$
$$CH$$
$$\|$$
$$CH_2$$
vinyl

High trans Polybutadiene   高反式聚丁二烯橡胶

  High trans BR is a <u>crystalline</u> plastic material similar to high trans   结晶的
<u>polyisoprene</u> or <u>balata</u>, which was used in golf ball covers. Notes below,   聚异戊二烯/巴拉塔胶
that in the <u>trans configuration</u> the main polymer chain is on opposite   反式构型
sides of the internal carbon-carbon double bond. Trans BR has a <u>melting</u>   熔点
<u>point</u> of about 80℃. It is made with transition metal catalysts similar to
the high cis process (La, Nd and Ni). These catalysts can make polymers
with >90% trans again using the <u>solution process</u>.   溶液聚合

$$\begin{array}{c} H \quad\quad CH_2\sim \\ \diagdown\ /\ \\ C=C \\ /\ \diagdown \\ \sim CH_2 \quad\quad H \end{array}$$
***trans*-1,4-**

New Words

butadiene	[ˌbjuːtəˈdaiiːn]	n.	丁二烯
polybutadiene	[ˌpɔliˌbjuːtəˈdaiiːn]	n.	聚丁二烯
tread	[tred]	n.	胎面
sidewall		n.	胎侧
impart	[imˈpɑːt]	vt.	给予，传授
fuel	[fjuəl]	n.	燃料
economy	[i(ː)ˈkɔnəmi]	n.	经济，节约
typically	[ˈtipikəli]	adv.	代表性地，主要地
traction	[ˈtrækʃən]	n.	牵引
impact	[ˈimpækt]	n.	冲击
modifier	[ˈmɔdifaiə]	n.	改性剂
polystyrene	[ˌpɔliˈstaiəriːn]	n.	聚苯乙烯(PS)
toughen	[ˈtʌfn]	v.	(使)变坚韧
resiliency	[riˈziliənsi,-jənsi]	n.	回弹性
alkyllithium			烷基锂
stereoregular	[ˌstiəriəˈregjulə]		立构规整的
backbone	[ˈbækbəun]	n.	主链
configuration	[kənˌfigjuˈreiʃən]	n.	构型
anionic	[ˌænaiˈɔnik]	adj.	阴离子的
thermal	[ˈθəːməl]	adj.	热的，热量的
stable	[ˈsteibl]	adj.	稳定的
pendent	[ˈpendənt]	adj.	悬挂的，侧基的

crystalline	[ˈkristəlain]	adj.	结晶的，晶态的
balata	[ˈbælətə]	n.	巴拉塔胶
lanthanum	[ˈlænθənəm]	n.	镧(La)
niobium	[naiˈəubiəm]	n.	铌(Nb)

Notes

[1] BR also has a major application as an impact modifier for polystyrene and acrylonitrile-butadiene-styrene resin (ABS) with about 25% of the total volume going into these applications. 另外大约有25%的顺丁橡胶用作聚苯乙烯和ABS树脂的抗冲改性剂。

[2] The transition metal, or so called Ziegler catalysts produce very "stereoregular" BRs with one type having the main polymer chain on the same side of the carbon-carbon double bond contained in the polybutadiene backbone. 采用过渡金属催化剂或称作齐格勒催化剂生产出的顺丁胶具有高的立构规整性，顺丁胶的分子长链位于主链中的碳-碳双键的同一侧。

Exercises

1. Translation

What is the "glass transition temperature"?

When different elastomers are being described, a fundamental property which is often quoted is the glass transition temperature, T_g, which differs from one elastomer to another. For example, for natural rubber T_g is −70℃. Broadly this means that above −70℃ the material behaves as a rubber, but below −70℃ the material behaves more like a glass. When glassy, natural rubber is about one thousand times as stiff as it is when rubbery. When glassy a hammer blow on natural rubber will cause it to shatter like a glass; when rubbery the hammer is likely just to bounce off.

Of course in practice the dividing line between the glassy and rubbery behaviour just described is not as sharp as this. In fact the transition is spread over some tens of degrees - but it is centred around −70℃. Thus, although a T_g can be accurately defined (although varying somewhat with the precise test conditions), for practical purposes we have to consider a glass transition region within which the properties are slowly changing from rubbery to glassy or vice versa (the processes are completely reversible). The broadness of this transition region varies from elastomer to elastomer.

2. Put the following words or expressions into Chinese

tread wear	rolling resistance	glass transition temperature
wet traction properties	tread compound	impact modifier
rubber-toughened resin	polymerization process	carbon-carbon double bond
trans configuration	transition metal catalyst	solution process
main chain	melting point	high trans polyisoprene

3. Translation

Mills

The speeds of the two rolls are often different. For natural rubber mixing a friction ratio of 1:1.25 for the front-to-back roll is common. High friction ratios are used for refining compounds, and even-speed rolls on feed mills to calenders. For mixing some of the synthetic rubbers, a near even-speed is best. Other synthetic rubbers are very difficult to mix on mills, so internal mixers are often used for these. Mills are fitted with a metal tray under the rolls to collect droppings from the mill, and with guides or cheeks, which are plates fitted to the ends of rolls to prevent rubber being contaminated with grease, etc.

[Reading Material]

Difference of NR and IR

Q: I understand that natural rubber and gutta percha are both polyisoprenes. Why then is one a rubber and the other a plastics-like material?

A: The difference lies in the case of crystallization. NR is the cis form of 1,4-polyisoprene and at ambient temperatures is seldom found in a crystalline state unless advantage is being taken of its ability to strain-crystallize - or self-reinforce itself on stretching. Raw NR can certainly crystallize during storage and harden, but the melting point never exceeds 30℃ and vulcanization (especially with sulphur) seriously inhibits the process.

Gutta percha, the trans form of polyisoprene, crystallizes very rapidly and has a melting point of between 65℃ and 75℃ - not high in comparison to most plastics, but ideal for many moulding applications. One other difference is that gutta has a much lower melt viscosity than rubber and flows readily once crystallites are removed - so "mould definition" can be very high.

Interestingly if gutta percha is vulcanized it is transformed into a rubber.

Words and Expressions

gutta percha			古塔波胶
crystalline state			晶态
take advantage of			利用
strain-crystallize			应变(诱导)结晶
self-reinforce			自补强
stretching	['stretʃiŋ]		拉伸, 伸长[展]
melting point			熔点
exceed	[ik'siːd]	vt.	超越, 胜过
comparison	[kəm'pærisn]	n.	比较, 对照
in comparison with (to)		adv.	与…比较
melt viscosity			熔体黏度

Unit 3　E-SBR

Types of SBR

There is a large variety of E-SBR types based on the styrene content, polymerization temperature, staining or non-staining antioxidants, oil and carbon black content. Each of these basic classifications include a variety of SBR polymer variations with respect to Mooney viscosities, coagulation types, emulsifier type, oil levels, and carbon black types and levels. Table 1 shows the basic groups of E-SBR.

Table 1　Numbering System for E-SBR

Series	Comments
1000	Hot polymerized polymers.
1500	Non-extended cold polymerized polymers.
1600	Non-oil-extended cold carbon black masterbatches
1700	Cold oil-extended polymers.
1800	Cold oil-extended carbon black masterbaches
1900	Miscellaneous high styrene resin masterbatches

Properties of E-SBR

Mechanical Properties

Since SBR lacks the self-reinforcing qualities of natural rubber due to stress induced crystallization, gum vulcanizates of SBR have lower tensile properties. The tensile property of E-SBR vulcanizates depends in great measure on the type and amount of filler in the compound. [1]Cured gum stocks have only 2.8 to 4.2 MPa tensile strength, while fine particle carbon black loadings can produce tensile strength of 27.6MPa. Though the compression set of some of the common E-SBR compounds is high, by proper compounding and blending, it is possible to obtain E-SBR vulcanizates with a low compression set.

Electrical Properties

SBR is a non-polar polymer and its vulcanizates are poor conductors of electricity. The electrical properties of E-SBR depend to a large extent on the amount and type of emulsifier and coagulating agent(s) used.

Resistance to Fluids

[2] While E-SBR vulcanizates are resistant to many polar solvents such as dilute acids and bases, they will swell considerably when in contact with gasoline, oils, or fats. Due to this limitation, SBR cannot be

used in applications that require resistance to swelling in contact with hydrocarbon solvents.

Cure Properties

SBR can be cured with a variety of cure systems including sulfur (accelerators and sulfur), peroxides and phenolic resins. Processing of SBR compounds can be performed in a mill, internal mixers or mixing extruders. SBR compounds are cured in a variety of ways by compression, injection molding, hot air or steam autoclaves, hot air ovens, microwave ovens and combinations of these techniques.

烃类溶剂
硫化性能
硫化体系
过氧化物/酚醛树脂
开炼机/密炼机/混炼用挤出机
模压硫化/注射硫化/硫化罐/烘箱/微波室

New Words

quality	['kwɔliti]	n.	质量，性质
gum	[gʌm]	n.	纯胶
stock	[stɔk]	n.	胶料
loading	['ləudiŋ]	n.	加入量
conductor	[kən'dʌktə]	n.	导体
emulsifier	[i'mʌlsifaiə]	n.	乳化剂
dilute	[dai'lju:t, di'l-]	adj.	淡的，稀释的
base	['beis]	n.	碱
swell	[swel]	v.	溶胀，增大
considerably	[kən'sɪdərəblɪ]	adv.	相当地
compression	[kəm'preʃ(ə)n]	n.	模压
injection	[in'dʒekʃən]	n.	注射
steam	[sti:m]	n.	蒸汽
autoclave	['ɔ:təukleiv]	n.	硫化罐
microwave	['maikrəuweiv]	n.	微波
technique	[tek'ni:k]	n.	技巧，方法
coagulation	[kəuˌægju'leiʃən]	n.	凝聚
extend	[iks'tend]	v.	填充
masterbatch			母炼胶
miscellaneous	[misi'leinjəs, -niəs]	adj.	混杂的

Notes

[1] Cured gum stocks have only 2.8 to 4.2 MPa tensile strength, while fine particle carbon black loadings can produce tensile strength of 27.6 MPa. 丁苯胶的纯胶硫化胶的抗张强度只有2.8～4.2 MPa，但当加入细粒子炭黑后抗张强度可提高到27.6 MPa。

[2] While E-SBR vulcanizates are resistant to many polar solvents such as dilute acids and bases, they will swell considerably when in contact with gasoline, oils, or fats. 尽管乳聚丁苯能耐许多极性溶剂（如稀酸和稀碱），但在汽油、石油系油类或脂肪烃类油中则会发生明显溶胀。

Exercises

1. Translate the following into Chinese

SBR (Styrene Butadiene Rubber)

A synthetic rubber which is easy to process in large quantities. Widely used in the footwear and tyre industries.

Properties
- good physical strength
- good tear and abrasion resistance
- range of colours
- one of the cheaper rubbers

Limitations
- does not resist oil or fuels
- prone to weathering

Typical Applications

Non-mechanical high-volume products such as shoe soles and heels or car tyres

2. Match column A with column B

Column A	Column B
生胶	vulcanizate
天然胶	gree stock
合成胶	oil-extended rubber
再生胶	masterbatch
并用胶	raw rubber
塑炼胶	reclaimed rubber
混炼胶	black-extended rubber
硫化胶	synthetic rubber
未硫化胶	compounded rubber
充炭黑胶	natural rubber
充油胶	masticated rubber
母炼胶	rubber blend

3. Translation

SBR Compounds

All types of SBR use the same basic compounding recipes as do other unsaturated hydrocarbon polymers. They need sulphur, accelerators antioxidants (and antiozonants), activators, fillers, and softeners or extenders. SBR requires less sulphur than natural rubber for curing. The usual range is about 1.5 ~ 2.0 phr of sulphur.

Compounding recipes with low sulphur improve aging properties, but are slower curing. Zinc stearate is the most common activator for SBR. There are many accelerators which are useful to speed up slow-curing stocks and retarders for slowing down the cure rate of scorchy stocks. Recipes may also contain plasticizers, tackifiers, softeners, waxes, reclaim, etc. It is not unusual to find a recipe with 15 or more ingredients.

Blends of SBR and other rubbers such as natural rubber or cis-polybutadiene are made for many uses.

[Reading Material]

Polymerization of E-SBR

As the molecular weight of the SBR increases, the vulcanizate resilience and the mechanical properties, particularly tensile strength and compression set, improve. The processability of SBR improves as its molecular weight distribution broadens. Formation of high molecular weight fractions with the increase in the average molecular weight can however, prevent improvements in the processability. This is due to the fact that the tendency for gel formation also increases at higher molecular weights.

In addition to the polymer viscosity, polymerization temperature also plays an important role in shaping the processability. E-SBRs produced at low polymerization temperatures have less chain branching than those produced at higher temperature. At an equivalent viscosity, cold polymerized E-SBR is normally easier to process than hot polymerized E-SBR, and this applies particularly to a better banding on mills, less shrinkage after calendering, and a superior surface of green tire compounds. Hot rubbers give better green strength because they have more chain branching.

The styrene content of most emulsion SBR varies from 0 to 50%. The percent styrene of most commercially available grades of E-SBR is 23.5%. In vulcanizates of SBR, as styrene content increases, dynamic properties and abrasion resistance decrease while traction and hardness increase.

Words and Expressions

processability			加工性
distribution	[ˌdistriˈbjuːʃən]	n.	分布
molecular weight distribution (MWD)			分子量分布
broaden	[ˈbrɔːdn]	v.	放宽，变宽，加宽
formation	[fɔːˈmeiʃən]	n.	形成，构成
fraction	[ˈfrækʃən]	n.	级分，组分
gel	[dʒel]	n.	凝胶
band	[bænd]	v.	包辊
shrinkage	[ˈʃrinkidʒ]	n.	收缩
calender	[ˈkælində]	n.	压延
superior	[sjuːˈpiəriə]	adj.	较高的，上级的
green tire			生胎，胎胚
green strength			生胶强度

Unit 4 Acrylonitrile-Butadiene Rubber

Properties and Applications

Nitrile Rubber (NBR) is commonly considered the workhorse of the industrial and automotive rubber products industries. NBR is actually a complex family of unsaturated copolymers of acrylonitrile and butadiene. [1]By selecting an elastomer with the appropriate acrylonitrile content in balance with other properties, the rubber compounder can use NBR in a wide variety of application areas requiring oil, fuel, and chemical resistance. In the automotive area, NBR is used in fuel and oil handling hose, seals and grommets. With a temperature range of − 40℃ to +125℃, NBR materials can withstand all but the most severe automotive applications. On the industrial side NBR finds uses in roll covers, hydraulic hoses, conveyor belting, oil field packers, and seals for all kinds of plumbing and appliance applications. Worldwide consumption of NBR is expected to reach 368000 metric tons annually by the year 2005.

Like most unsaturated thermoset elastomers, NBR requires formulating with added ingredients, and further processing to make useful articles. Additional ingredients typically include reinforcement fillers, plasticizers, protectants, and vulcanization packages. Processing includes mixing, pre-forming to required shape, application to substrates extrusion, and vulcanization to make the finished rubber article. Mixing and processing are typically performed on open mills, internal mixers, extruders, and calenders. Finished products are found in the marketplace as injection or transfer molded products (seals and grommets), extruded hose or tubing, calendered sheet goods (floor mats and industrial belting), or various sponge articles.

Acrylonitrile (ACN) Content

The ACN content is one of two primary criteria defining each specific NBR grade. [2]The ACN level, by reason of polarity, determines several basic properties, such as oil and solvent resistance, low-temperature flexibility/glass transition temperature, and abrasion resistance. Higher ACN content provides improved solvent, oil and abrasion resistance, along with higher glass transition temperature. Table 1 summarizes most of the common properties for conventional NBR polymers. The direction of the arrows signifies an increase/improvement in the values.

Table 1 NBR Properties – Relationship to Acrylonitrile Content

NBR with Lower Acrylonitrile Content		NBR with Higher Acrylonitrile Content	
Processability	→		
Cure Rate w/Sulfur Cure System	→		硫化速率/硫黄硫化体系
Oil / Fuel Resistance	→		
Compatibility w/Polar Polymers	→		相容性
Air / Gas Impermeability	→		气密性
Tensile Strength	→		
Abrasion Resistance	→		
Heat-Aging	→		
	←	Cure Rate w/Peroxide Cure System	过氧化物硫化体系
	←	Compression Set	
	←	Resilience	
	←	Hysteresis	滞后
	←	Low Temperature Flexibility	

New Words

workhorse	[ˈwəːkhɔːs]	n.	主力军
compounder		n.	配料员，配方人员
grommet	[ˈgrɔmit]	n.	垫圈
withstand	[wiðˈstænd]	vt.	抵挡，经受住
all but			几乎，差一点
hydraulic	[haiˈdrɔːlik]	adj.	液压的，水压的
packer	[ˈpækə]	n.	垫片
plumbing	[ˈplʌmiŋ]	n.	管
appliance	[əˈplaiəns]	n.	用具，器具
unsaturated	[ˈʌnˈsætʃəreitid]	adj.	不饱和的
substrate	[ˈsʌbstreit]	n.	底层，基层，被粘物
extruder	[eksˈtruːdə]	n.	压出机
compatibility	[kəmˌpætiˈbiliti]	n.	相容性
polarity	[pəuˈlæriti]	n.	极性
summarize	[ˈsʌməraiz]	v.	概述，总结
signify	[ˈsignifai]	vt.	表示，意味
hysteresis	[ˌhistəˈriːsis]	n.	滞后作用

Notes

[1] By selecting an elastomer with the appropriate acrylonitrile content in balance with other properties, the rubber compounder can use NBR in a wide variety of application areas requiring oil, fuel, and chemical resistance. 只要丙烯腈的含量选择适当并综合考虑到其他性能，丁腈胶可以用在大多数要求耐油、耐化学腐蚀的制品中。

[2] The ACN level, by reason of polarity, determines several basic properties, such as oil and solvent resistance, low-temperature flexibility/glass transition temperature, and abrasion

resistance. 由于丙烯腈是极性的，它的含量决定了丁腈胶的一些基本性能，如耐油性、耐溶剂性、低温柔韧性（玻璃化温度）和耐磨性。

Exercises

1. Translate the following into Chinese

Nitrile (Acrylonitrilebutadiene)

An early development in the search for an oil resistant rubber. The most suitable rubber for applications requiring resistance to petroleum based fluids (there are rubbers with higher degrees of resistance but these are much more expensive).

Properties
- very good resistance to petroleum based fluids
- good high temperature resistance - up to 100℃ (120℃ with EV cure systems)
- economical to compound and produce
- very low level of permeability to gases

Limitations
- poor resistance to outdoor weathering without special compounding
- comparatively low strength
- flammable and burns with toxic fumes

Typical Applications
- sealing in enclosed spaces where there is contact with petroleum based fluids
- sealing against gases

2. Put the following words or expressions into Chinese

roll cover	conveyor belting	thermoset elastomer
reinforcement filler	open mill	internal mixer
calender	extruder	sponge article
acrylonitrile content	polar polymer	gas impermeability
hydraulic hose	cure rate	transfer molded product

3. Put the following compounds used in rubber industry into English

乙烯	丙烯	丁二烯	苯乙烯	异戊二烯
氯丁二烯	氧化锌	丙烯腈	硫黄	异丁烯
丙烯酸酯	硬脂酸	硬脂酸锌	秋兰姆	
顺式 1,4-聚丁二烯			乙烯-丙烯共聚物	

[Reading Material]

General Types of NBR

Cold NBR

The current generation of cold NBR's spans a wide variety of compositions. Acrylonitrile content ranges from 15% to 51%. Mooney values range from a very tough 110, to pourable liquids, with 20 ~ 25 as the lowest practical limit for solid material. They are made with a wide array of emulsifier systems, coagulants, stabilizers, molecular weight modifiers, and chemical compositions.

Third monomers are added to the polymer backbone to provide advanced performance. Each variation provides a specific function.

Cold polymers are polymerized at a temperature range of 5℃ to 15℃, depending on the balance of linear-to-branched configuration desired. The lower polymerization temperatures yield more-linear polymer chains. Reactions are conducted in processes universally known as continuous, semi-continuous and batch polymerization. Figure1 shows the chemical structure of NBR, indicating the three possible isomeric structures for the butadiene segments.

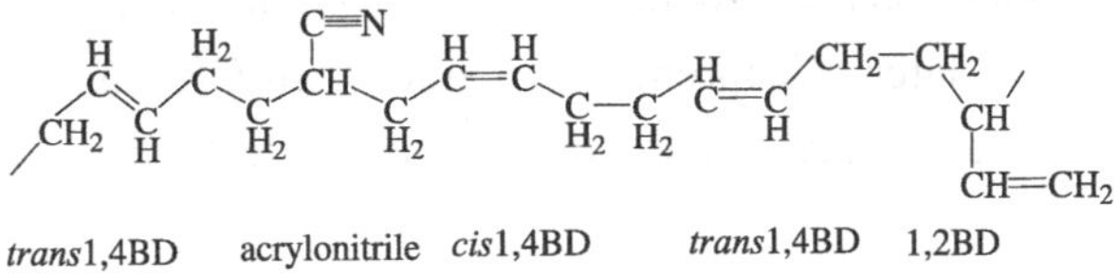

*trans*1,4BD   acrylonitrile   *cis*1,4BD   *trans*1,4BD   1,2BD

Figure 1 NBR Strucutre

Carboxylated Nitrile (XNBR)

Addition of carboxylic acid groups to the NBR polymer's backbone significantly alters processing and cured properties. The result is a polymer matrix with significantly increased strength, measured by improved tensile, tear, modulus and abrasion resistance. The negative effects include reduction in compression set, water resistance, resilience and some low-temperature properties.

Words and Expressions

span	[spæn]	n.	跨度，范围
pourable	[ˈpɔːrəbl]	adj.	可倾倒的
array	[əˈrei]	n.	大批
coagulant	[kəʊˈægjʊlənt]	n.	凝聚剂
molecular weight modifier			分子量调节剂
universally	[juːniˈvɜːsəli]	adv.	普遍地，全体地，到处
batch	[bætʃ]	n.	间歇
isomeric	[ˌaisəuˈmerik]	adj.	异构的
carboxylate	[kɑːˈbɔksileit]	n.	羧酸盐，羧酸酯
carboxylic	[ˌkɑːbɔkˈsilik]	adj.	羧基的
carboxylic acid			羧酸

Unit 5　Chloroprene Rubber

氯丁胶

Applications

Among the speciality elastomers polychloroprene [poly(2-chloro-1,3-butadiene)] is one of the most important with an annual consumption of nearly 300000 tons worldwide. First production was in 1932 by DuPont ("Duprene", later "Neoprene") and since then CR has an outstanding position due to its favourable combination of technical properties.

特种胶

工艺性能

[1] CR is used in different technical areas, mainly in the rubber industry (ca.61%), but is also important as a raw material for adhesives (both solvent based and water based, ca.33%) and has different latex applications (ca.6%) such as dipped articles (e.g. gloves), moulded foam and improvement of bitumen.

溶剂基/水基（胶黏剂）
胶乳/浸渍制品
泡沫

Rubber Properties

CR is not characterised by one outstanding property, but its balance of properties is unique among the synthetic elastomers. It has:
- Good mechanical strength
- High ozone and weather resistance
- Good aging resistance
- Low flammability
- Good resistance toward chemicals
- Moderate oil and fuel resistance
- Adhesion to many substrates

耐老化
阻燃性
耐化学药品性
耐油性

Polychloroprene can be vulcanized by using various accelerator systems over a wide temperature range.

Types

Normal linear grades (general-purpose grades):

通用型氯丁胶

General-purpose grades are mostly produced with n-dodecyl mercaptan as the chain transfer agent and occasionally with xanthogen disulfides. If xanthogen disulfides are used, the elastomers are more readily processible and give vulcanizates with improved mechanical properties.

正十二碳硫醇
链转移剂
二硫化黄原酸酯

Precrosslinked grades:

预交联型氯丁胶

Precrosslinked grades consist of a blend of soluble polychloroprene and crosslinked polychloroprene. They show less swelling after extrusion (die swell) and better calenderability. Precrosslinked grades are particularly suitable for the extrusion of profiled parts.

共混物
交联聚氯丁二烯
挤出胀大/压延性

Sulfur-modified grades: | 硫调型氯丁胶

Sulfur-modified grades are copolymers of chloroprene and elemental sulfur. [2]The viscosity is adjusted – in contrast to general-purpose grades – mostly after polymerization by "peptization" of the polysulfide bonds by e.g. thiuram disulfides as peptization agents. Sulfur modification improves the breakdown of the rubber during mastication (lowering of viscosity). [3]Sulfur-modified grades are used in particular for parts exposed to dynamic stress, such as driving belts, timing belts or conveyor belts because of their excellent mechanical properties. But the polymers are less stable during storage and the vulcanizates less resistant to aging.

单质硫

多硫键/二硫化秋兰姆/调节剂/断链

塑炼

动态应力/传动带
同步带/运输带

Slow cristallizing grades: | 低结晶性氯丁胶

Slow cristallizing grades are polymerized with 2,3-dichloro-1,3-butadiene as a comonomer. [4]This comonomer reduces the degree of cristallization by introducing irregularities into the polymer chain. High polymerization temperatures also make structural irregularities, if this comonomer is not available. Crystallization resistant grades are used to produce rubber articles, which have to retain their rubbery properties at very low temperatures.

共聚单体

结晶度

New words

ca. (=about)			大约
bitumen	['bitjumin]	n.	沥青
flammability	[,flæmə'biləti]	n.	易燃，可燃性
dodecyl	['dəudəsil]		十二(烷)基
mercaptan	[mə'kæptæn]	n.	硫醇
xanthogen	['zænθəudʒin]		黄原酸
processible	['prəusesəbl; `prɔ-]	adj.	可加工的
precrosslinked			预交联的
calenderability	[,kælindərə'biliti]		压延性能
profiled part			异型配件
in contrast to			和…形成对比
peptization	[,peptai'zeiʃən]		塑解
breakdown		n.	破坏
polymerize	['pɔliməraiz]	v.	(使)聚合
2,3-dichloro-1,3-butadiene			2,3-二氯-1,3-丁二烯
comonomer	[kəu'mɔnəmə]	n.	共聚单体
irregularity	[i,regju'læriti]	n.	不规则，无规律

Notes

[1] CR is used in different technical areas, mainly in the rubber industry (ca.61%), but is also important as a raw material for adhesives (both solvent based and water based, ca.33%) and has different latex applications (ca.6%) such as dipped articles (e.g. gloves), moulded foam and improvement of bitumen. 氯丁胶用在不同的技术领域，主要用在橡胶工业中（约占 61%），也是胶黏剂的重要原料（既可制成溶剂基也可制成水基胶黏剂，约占 33%），另有一些以氯丁胶乳形式使用，如浸渍制品（乳胶手套），模型泡沫制品和用作沥青的改性剂（约占 6%）。

[2] The viscosity is adjusted – in contrast to general-purpose grades - mostly after polymerization by "peptization" of the polysulfide bonds by e.g. thiuramdisulfides as peptization agents. 与通用型 CR 不同的，硫黄调节型 CR 的黏度可调，大多是在聚合后通过破坏多硫键（如二硫化秋兰姆类化合物作为塑解剂）来调节黏度。

[3] Sulfur-modified grades are used in particular for parts exposed to dynamic stress, such as driving belts, timing belts or conveyor belts because of their excellent mechanical properties. 硫黄调节型氯丁胶由于力学性能优异特别用作动态条件下使用的橡胶制品，如传动带、同步齿形带或运输带。

[4] Slow cristallizing grades are polymerized with 2,3-dichloro-1,3-butadiene as a comonomer. This comonomer reduces the degree of cristallization by introducing irregularities into the polymer chain. 难结晶型氯丁胶是与 2,3-二氯-1,3-丁二烯共聚而成的氯丁胶，共聚单体破坏了分子链的规整性从而使结晶度减小。

Exercises

1. Translate the following into Chinese

Neoprene(Polychloroprene)

One of the first synthetic rubbers developed in the search for oil resistant rubber. Widely used due to its combination of useful properties and comparatively low price.

Properties
- resistant to a wide range of hostile environments
- resistant to oils and chemicals
- weather and water resistant
- can withstand temperatures from −30℃ to 95℃
- easy to process and compound, offering cost benefits
- flame retardant
- can be produced in any colour required

Limitations
- unsuitable for applications requiring contact with fuels
- tendency to tear once there is initial damage

some Neoprenes （Neoprene is a registered trade mark of Du Pont.） may crystallise during storage or use causing temporary stiffening (increase in modulus/hardness). If parts are deformed during crystallisation, they may take on a set. However, crystallisation is a readily reversible *phenomenon and can be removed by warming over 80 ℃. It can be prevented by the use of special grades.*

Typical Applications
- most general mechanical applications without contact with fuel
- particularly useful in marine environments due to good ozone resistance.

2. Put the following words or expressions into Chinese

speciality elastomer	mechanical strength	die swell	2,3-dichloro-1,3-butadiene
annual consumption	ozone resistance	calenderability	degree of cristallization
technical properties	weather resistance	profiled part	polymer chain
technical area	aging resistance	flammability	structural irregularity
rubber industry	precrosslinked grade	peptization agent	rubber article
oil resistance	sulfur-modified grade	driving belt	solvent based adhesive
dipped article	*n*-dodecyl mercaptan	timing belt	resistance to chemicals
chain transfer agent	mechanical properties	conveyor belt	swelling after extrusion
xanthogen disulfide	water based adhesive	aging resistance	polysulfide bond

3. Put the following words or expressions into English

物理性能	化学性能	力学性能	加工性能
工艺性能	屈挠性能	老化性能	拉伸性能
耐磨性	耐化学药品性	拉伸强度	伸长率
永久变形	耐油性	动态性能	耐候性
电性能	硫化性能	撕裂强度	气密性
硬度	定伸强度	蠕变性能	耐臭氧

[Reading Material]

Rubber Compounding Rules (1)

The compounding rules to achieve the properties wanted are shortly given below.

Hardness

Hardness and reinforcement are determined by the amount and the type of filler, by the degree of dispersion and by the cross-link density. Carbon black is the most common rubber filler in engineering applications. Sometimes white fillers such as silica and clay are used instead of carbon black. For maximum tensile properties about 25 volume percent carbon black is needed. To lower the hardness at that level softeners are used.

For low creep properties the level of carbon black should be kept to the minimum acceptable level. If the application requires a high resistance to abrasion a small particle size type carbon black is necessary. The stiffness of a rubber product is determined by result of the modulus of the rubber, the mode of deformation, the shape of the product and its dimensions.

Modulus

The modulus of a rubber is also determined by the amount of filler. For a low modulus product normally low reinforcing blacks or (non-reinforcing) white fillers are used. The modulus can be influenced by the rubber grade (viscosity) and by the vulcanization system.

Low modulus properties could be reached by the use of a soluble efficient vulcanization (EV)-system, based on soluble accelerators, activators and low sulfur level. The modulus of black filled natural rubber is almost independent of the temperature over a range from about -20℃ to over +100℃. Below -20℃ the modulus increases as the temperature is lowered. The addition of 20 parts of di-iso-octyl sebacate (DOS) per hundred parts of natural rubber will lower the temperature performance by about 10℃.

High resilience

Resilience decreases as filler level is increased. For high resilience the filler level has to be kept to an acceptable amount to maintain (physical) properties. Medium particle size/high structure blacks give low hysteresis because only moderate amounts are needed to increase modulus.

High damping

High damping can be reached by filler/ oil extensions using high viscosity oil or by blending with synthetic rubbers such as SBR, EPDM or NIR. The damping is a consequence of the synthetic polymer having a T_g not far under normal service temperatures. A further consequence of the proximity of the T_g to room temperature is that the modulus of the blend increases significantly below about 10℃ especially when the proportion of the acrylonitrile /isoprene rubber (NIR) is high. For this reason the most useful blends are probably covered by the range NR/NIR from 90/10 to 70/30.

Low compression set

Engineering products with low compression set values in outdoor applications can be vulcanized with organic peroxides. However, large compression mould products are prone to scorch because it is not possible to delay peroxide vulcanizion by using delayed action chemicals. Secondly peroxide vulcanized products have a poor tear strength and are incompatible with anti-ozonants. Therefore, in practice, conventional vulcanization (CV) or semi-efficient vulcanization (SEV)-systems are normally used.

Low creep/relaxation

For good creep/relaxation resistance the filler content has to be kept at a minimum acceptable level and uses medium particle sized blacks. Replacement of stearic acid by zinc-2- ethyl hexanoate (ZEH) lowers the physical creep/relaxation rate. For a low creep/relaxation rate a soluble Evsystem is advised. A relaxation rate of less than 3 % per decade of time is possible. By the use of the low sulfur system the long term behaviour of those products is very good. The good thermal stability keeps also the secondary chemical creep/relaxation and compression set at elevated temperatures comparatively low.

Words and Expressions

determine	[di'tə:min]	v.	决定, 确定
dispersion	[dis'pə:ʃən]	n.	分散
cross-link density			交联度
creep	[kri:p]	vi.	蠕变
modulus	['mɔdjuləs]	n.	定伸强度
mode	[məud]	n.	方式, 模式

soluble	['sɔljubl]	adj.	可溶的，可溶解的
diisooctyl	[dai'aisəuɔktəl]	n.	二异辛基
sebacate	['sebəkeit]	n.	癸二酸盐(或酯)
di-iso-octyl sebacate (DOS)			癸二酸二异辛酯
NIR(acrylonitrile-isoprene rubber)			丙烯腈–异戊二烯胶
proximity	[prɔk'simiti]	n.	接近
portion	['pɔ:ʃən]	n.	部分
prone to			倾向于…
scorch	[skɔ:tʃ]	n.	焦烧
delay	[di'lei]	v.	耽搁，延迟，
incompatible	[ˌinkəm'pætəbl]	adj.	不相容的
relaxation	[ˌri:læk'seiʃən]	n.	应力松弛
zinc-2- ethyl hexanoate (ZEH)			2–乙基己酯锌
long term			长期
secondary	['sekəndəri]	adj.	次要的，第二的
comparatively	[kəm'pærətɪvlɪ]	adv.	比较地，相当地

Unit 6 Butyl Rubber

Properties and Applications

Butyl rubber (IIR) is the copolymer of isobutylene and a small amount of isoprene. [1]First commercialized in 1943, the primary attributes of butyl rubber are excellent impermeability/air retention and good flex properties, resulting from low levels of unsaturation between long polyisobutylene segments. Tire innertubes were the first major use of butyl rubber, and this continues to be a significant market today.

[2] The development of halogenated butyl rubber (halobutyl) in the 1950's and 1960's greatly extended the usefulness of butyl by providing much higher curing rates and enabling covulcanization with general purpose rubbers such as natural rubber and styrene-butadiene rubber (SBR). These properties permitted development of more durable tubeless tires with the air retaining innerliner chemically bonded to the body of the tire. Tire innerliners are by far the largest application for halobutyl today. Both chlorinated (chlorobutyl) and brominated (bromobutyl) versions of halobutyl are commercially available. [3]In addition to tire applications, butyl and halobutyl rubbers' good impermeability, weathering resistance, ozone resistance, vibration dampening, and stability make them good materials for pharmaceutical stoppers, construction sealants, hoses, and mechanical goods.

Processing and Vulcanization

Like other rubbers, for most applications, butyl rubber must be compounded and vulcanized (chemically cross-linked) to yield useful, durable end use products. [4]Grades of Butyl have been developed to meet specific processing and property needs, and a range of molecular weights, unsaturation, and cure rates are commercially available. Both the end use attributes and the processing equipment are important in determining the right grade of butyl rubber to use in a specific application. The selection and ratios of the proper fillers, processing aids stabilizers, and curatives also play critical roles in both how the compound will process and how the end product will behave.

Elemental sulfur and organic accelerators are widely used to cross-link butyl rubber for many applications. The low level of unsaturation requires aggressive accelerators such as thiuram or thiocarbamates. [5]The vulcanization proceeds at the isoprene site with the polysulfidic crosslinks attached at the allylic positions, displacing the allylic hydrogen. The number of sulfur atoms per cross-link is between one and four or more. Cure rate and cure state (modulus) both increase if the diolefin content is increased (higher unsaturation). Sulfur cross-links have limited stability at sustained high temperature. Resin cure systems (commonly using alkyl phenol-formaldehyde derivatives) provide for carbon-carbon crosslinks and more stable compounds.

烯丙氢

二烯烃含量
树脂硫化体系
烷基酚醛树脂
碳碳交联键

In halobutyl, the allylic halogen allows easier cross-linking than does allylic hydrogen alone, because halogen is a better leaving group in nucleophilic substitution reactions. Zinc oxide is commonly used to cross-link halobutyl rubber, forming very stable carbon-carbon bonds by alkylation through dehydrohalogenation, with zinc chloride byproduct. Bromobutyl is faster curing than chlorobutyl and has better adhesion to high unsaturation rubbers. As a result, its volume growth rate has exceeded that of chlorobutyl in recent decades as tire plants have driven to higher productivity operation.

亲核取代反应

烷基化/脱卤化氢/氯化锌/副产物

轮胎厂

New Words

innertube		n.	内胎
halogenate	[ˈhælədʒəneit]	v.	卤化
durable	[ˈdjuərəbl]	adj.	持久的，耐用的
tubeless		adj.	无内胎的
innerliner		n.	气密层
chlorinate	[ˈklɔːrineit]	vt.	用氯(或氯化物)处理，氯化
chlorobutyl			氯化丁基胶
brominate	[ˈbrəumineit]	vt.	用溴(或溴化物)处理，使溴化
bromobutyl			溴化丁基胶
version	[ˈvəːʃən]	n.	类型
dampen	[ˈdæmpən]	v.	阻尼
pharmaceutical	[ˌfɑːməˈsjuːtikəl]	n.	药物
		adj.	制药(学)上的
stopper	[ˈstɔpə]	n.	塞子
sealant	[ˈsiːlənt]	n.	密封剂
critical	[ˈkritikəl]	adj.	关键的
aggressive	[əˈgresiv]	adj.	活性高的
thiocarbamate	[ˌθaiəuˈkɑːbəmeit]		硫代氨基甲酸盐
allylic	[ˈæləˌlik]	adj.	烯丙基的
displace	[disˈpleis]	vt.	移置，取代，置换
diolefin	[daiˈəuləfin]	n.	二烯
sustained	[səsˈteind]	adj.	持续不变的
halogen	[ˈhælədʒən]	n.	卤素
alkylation	[ˌælkiˈleiʃ(ə)n]	n.	烷基化

Notes

[1] First commercialized in 1943, the primary attributes of butyl rubber are excellent impermeability/air retention and good flex properties, resulting from low levels of unsaturation between long polyisobutylene segments. 丁基胶于1943年商品化，丁基胶的主要特性表现为气密性优异和耐屈挠性好，这是由于在长聚异丁烯链段间只含少量的不饱和结构。

[2] The development of halogenated butyl rubber (halobutyl) in the 1950's and 1960's greatly extended the usefulness of butyl by providing much higher curing rates and enabling covulcanization with general purpose rubbers such as natural rubber and styrene-butadiene rubber (SBR). 20世纪50~60年代发展起来的卤化丁基胶扩大了丁基胶的使用范围，因为卤化丁基的硫化速度快且能与通用胶如天然、丁苯共硫化。

[3] In addition to tire applications, butyl and halobutyl rubbers' good impermeability, weathering resistance, ozone resistance, vibration dampening, and stability make them good materials for pharmaceutical stoppers, construction sealants, hoses, and mechanical goods. 除了可用在轮胎工业中，丁基胶和卤化丁基胶的优异的气密性、耐候性、抗臭氧、减震和稳定性使得它们成为制作药物胶塞、建筑密封胶、胶管和其他橡胶工业制品的最合适的材料。

[4] Grades of Butyl have been developed to meet specific processing and property needs, and a range of molecular weights, unsaturation, and cure rates are commercially available. 已经生产出各种不同的分子量、饱和度和硫化速度的丁基胶以满足特定的加工和性能需求。

[5] The vulcanization proceeds at the isoprene site with the polysulfidic cross links attached at the allylic positions, displacing the allylic hydrogen. 丁基胶的硫化发生在异戊二烯单元上，烯丙基位置上的氢被取代形成多硫交联键。

Exercises

1. Translate the following into Chinese

Polyisoprene

Most types of synthetic rubber represent man's attempt to reproduce or improve upon the physical behavior of natural rubber (NR). Only one, synthetic polyisoprene (IR), approximates the chemical composition of natural rubber. It finds its largest application in tyres, mechanical goods, and footwear.

Polyisoprene is generally more economical to process, lighter in colour, more uniform, and of higher purity. On the other hand, IR is lower in green (unvulcanized) strength, most noticeably in reworked stocks. In general, the synthetic polyisoprenes are lower in modulus and higher in elongation than the natural product.

Commercial synthetic *cis*-1,4-polyisoprene is a light amber or water-white, essentially odourless elastomer. Except for oil-extended grades, isoprene rubber is almost 100% rubber hydrocarbon; it contains no fatty acids, resins, or proteins as are found in natural rubber (about 6%). Nonrubber constituents in IR consist chiefly of stabilizer, residual moisture and trace quantities of inorganic materials. At present all commercial types contain a nonstaining, nondiscolouring stabilizer. The specific gravity of synthetic polyisoprene is 0.90~0.91.

In general, synthetic polyisoprene behaves like natural rubber during mixing and subsequent processing steps. Since the two polymers are so similar chemically, similar processability would be expected. Thus, both IR and NR show a relatively rapid decrease in viscosity with mastication compared to most other elastomers. Both polymers undergo predominantly chain scissionduring mixing and subsequent operations, with a corresponding reduction in molecular weight. At high temperatures (>175℉) in the presence of air, both polymers break down primarily by oxidative degradation. At room temperature the main mechanism of molecular weight reduction is mechanical chain scission. Neither IR nor NR changes in microstructure during normal processing.

The time to incorporate carbon black or other fillers is about the same as with premasticated NR. Polyisoprene tends to become sticky and lose green strength if overmixed or subjected to prolonged milling.

2. Put the following words or expressions into Chinese

impermeability to air   air retention   flex properties   curing rate
pharmaceutical stopper   resin cure system   allylic hydrogen   processing aid
aggressive accelerator   mechanical goods   diolefin content   tubeless tire
nucleophilic substitution reaction   zinc oxide   zinc chloride   tire plant

3. Put the following words or expressions into English

丁基胶   卤化丁基胶   氯化丁基胶   溴化丁基胶
异丁烯   异戊二烯   共聚物   共硫化
链段   稳定剂   多硫键   碳碳键
卤素   氧化锌   氯化锌   气密层

[Reading Material]

Rubber Compounding Rules (2)

Heat resistance

Heat resistance (including resistance to compression set, creep and stress relaxation at high temperatures) is performed by peroxide vulcanization. It gives the best possible thermal stability plus low creep rate, however, those vulcanizates have inferior mechanical properties. To attain the best heat protection, peroxide vulcanization must be carried out to completion (eg. post-cure) since un-reacted peroxide acts as a pro-oxidant. To overcome inferior mechanical properties, vulcanization based on soluble EV system could be used in conjuction with most powerful antidegradants.

By using anti-degradants a mixture of anti-oxidants is advisable, because mixtures are more effective than a single anti-oxidant. For nonblack compounds phenolic antioxidants are used because their non-staining properties. There is evidence that for protection against high temperature ageing it is best to use an anti-oxidant that has a low volatility even though it may be less chemically active than a more volatile one.

Cross-link density

The degree of cross-linking affects various properties. The cross-link density for maximum tear strength is slightly lower than for maximum tensile strength and abrasion. Resilience, compression set, creep and relaxation resistance are best at somewhat higher levels of cross-linking(Figure1).

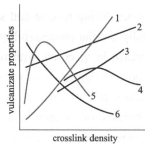

Figure 1   Vulcanizate properties vs. crosslink density

1. Static modulus, extrusion resistance.
2. High speed dynamic modulus.
3. Hardness.
4. Tensile strength.
5. Tear strength, fatigue life and toughness.
6. Hysteresis, permanent set, coefficient of friction, elongation and compression set.

Weathering resistance

Weathering of rubber is mainly degradation near the surface that can be attributed to the effects of oxygen, ozone and ultra-violet light. Engineering components made of rubber are seldom positioned in direct sunlight. UV-light is in rubbers for outdoor applications mainly no problem because carbon black in rubber acts as a UV-absorber.

For non-black compounds phenolic anti-oxidants are used or UVstabilizers. In most cases, traces of ozone attack the rubber surface when the rubber is stretched. Therefore it is advisable to include waxes or chemical anti-ozonants. For the use under static strains hydrocarbon waxes will prevent formation of cracks, provided that a wax appropriate to the exposure temperature is chosen. For dynamic strains an anti-ozonant of the substituted para-phenylene diamine class, either alone or with wax, is necessary. In special occasions it is possible to blend natural rubber with an amount of 25 phr EPDM rubber based on polymer.

Flex-cracking resistance

Fatigue life or flex-cracking resistance is highly sensitive to strain, decreasing as the maximum strain is increased. Components should be designed to operate, where possible, under static load over which the dynamic load is superimposed. For good flex cracking resistance the use of para-phenylene diamine anti-ozonants is advisable. In particular alkyl-aryl derivatives are effective, they also provide protection against oxidation.

Crystallization

Crystallization occurs when rubber is stressed or frozen. When rubber is stressed crystallization begins at moderate strains and at higher stresses more crystallites are formed. The reason for this phenomenon is that the crystallites are oriented in one direction of the extension. The crystalline structure is primarily responsible for the high strength and tear resistance. At low temperatures ($-25\,^\circ\mathrm{C}$) the rubber crystallize due to freezing. This effect is a reversible process, when temperature rises the formed crystals melt quickly. Vulcanized rubber crystallizes more slowly than raw rubber due to crosslinking. CV-systems (high sulfur systems) are extremely resistant to low temperature crystallization.

Bonding

The techniques for bonding are well known in the industry. For high bond strength the use of CV-systems gives the best results. For large products (bridge bearings) with a large surface area, semi-EV or EV systems could be used. In case of low temperature applications EV-systems crystallize rapidly, in that case high sulfur systems (CV) give products that are resistant to low temperatures. To obtain satisfactory bonds, the steel inlay's have to be well cleaned and avoid bloom on the rubber surface in the case of compression moulding.

Words and Expressions

vs. (versus)	[ˈvəːsəs]	prep.	对
static modulus			静态模量
dynamic	[daiˈnæmik]	adj.	动态的
inferior	[inˈfiəriə]	adj.	差的，次的
post-cure			二次硫化
pro-oxidant			助氧化剂
in conjunction with		adv.	与…协力
volatility	[ˌvɔləˈtiliti]	n.	挥发性
position	[pəˈziʃən]	vt.	安置
UV-absorber			紫外线吸收剂
provided	[prəːˈvaidid]	conj.	倘若
appropriate to			适于，合乎
occasion	[əˈkeiʒən]	n.	场合，时机，机会
superimpose	[ˈsjuːpərimˈpəuz]	v.	添加，双重
alkyl-aryl derivative			烷基-芳基衍生物
freeze	[friːz]		(使)结冰，(使)冷冻
crystallite	[ˈkristəlait]	n.	微晶
orient	[ˈɔːrient]	v.	取向，定向
reversible	[riˈvəːsəbl]	adj.	可逆的
bonding	[ˈbɔndiŋ]		粘接，接合
bond strength			黏结强度
inlay	[ˈinˈlei]	n.	镶嵌

Unit 7 Ethylene-Propylene Rubber

Properties and Applications

[1] Ethylene-propylene rubbers & elastomers (also called EPDM and EPM) continue to be one of the most widely used and fastest growing synthetic rubbers having both specialty and general-purpose applications. Sales have grown to 870 metric tons in 2000 since commercial introduction in the early 1960's. Versatility in polymer design and performance has resulted in broad usage in automotive weather-stripping and seals, glass-run channel, radiator, garden and appliance hose, tubing, belts, electrical insulation, roofing membrane, rubber mechanical goods, plastic impact modification, thermoplastic vulcanizates and motor oil additive applications.

[2] Ethylene-propylene rubbers are valuable for their excellent resistance to heat, oxidation, ozone and weather aging due to their stable saturated polymer backbone structure. Properly pigmented black and non-black compounds are color stable. As non-polar elastomers, they have good electrical resistivity, as well as resistance to polar solvents, such as water, acids, alkalies, phosphate esters and many ketones and alcohols. Amorphous or low crystalline grades have excellent low temperature flexibility with glass transition points of about minus 60℃.
[3] Heat aging resistance up to 130℃ can be obtained with properly selected sulfur acceleration systems and heat resistance at 160℃ can be obtained with peroxide cured compounds. Compression set resistance is good, particularly at high temperatures, if sulfur donor or peroxide cure systems are used.

These polymers respond well to high filler and plasticizer loading, providing economical compounds. They can develop high tensile and tear properties, excellent abrasion resistance, as well as improved oil swell resistance and flame retardance. A general summary of properties is shown in Table 1.

Processing and Vulcanization

The processing, vulcanization and physical properties of ethylene-propylene elastomers are largely controlled by the characteristics of ethylene content, diene content, molecular weight (or Mooney viscosity) and molecular weight distribution. For example, decreasing ethylene content decreases crystallinity and associated properties such as hardness and modulus.

Table 1 Properties of Ethylene-Propylene Elastomers

Polymer Properties	
Mooney Viscosity, ML 1+4 @ 125℃	5 ~ 200
Ethylene Content, wt. %	45 ~ 80
Diene Content, wt. %	0 ~ 15
Specific Gravity, g/ml	0.855 ~ 0.88
Vulcanizate Properties	
Hardness, <u>Shore A Durometer</u>	30 ~ 95
Tensile Strength, MPa	7 ~ 21
Elongation, %	100 ~ 600
Compression Set B, %	20 ~ 60
Useful Temperature Range, ℃	-50° ~ 160°
Tear Resistance	Fair ~ Good
Abrasion Resistance	Good ~ Excellent
Resilience	Fair ~ Good
Electrical Properties	Excellent

邵氏 A 硬度计

New words

demanding	[diˈmɑːndiŋ; diˈmændiŋ]	adj.	过分要求的, 苛求的
radiator	[ˈreidieitə]	n.	散热器, 水箱
membrane	[ˈmembrein]	n.	膜, 隔膜
resistivity	[ˌriːzisˈtiviti]	n.	抵抗力, 电阻
ester	[ˈestə]	n.	酯
durometer	[djuəˈrɔmitə]	n.	硬度计
elongation	[ˌiːlɔŋˈgeiʃən]	n.	伸长

Notes

[1] Ethylene-propylene rubbers & elastomers (also called EPDM and EPM) continue to be one of the most widely used and fastest growing synthetic rubbers having both specialty and general-purpose applications. 乙丙胶(EPM 和 EPDM)一直是应用最广泛和发展最快的合成胶之一, 它既能用作通用胶又能用作特种胶。

[2] Ethylene-propylene rubbers are valuable for their excellent resistance to heat, oxidation, ozone and weather aging due to their stable, saturated polymer backbone structure. 乙丙胶是主链饱和橡胶, 性质稳定, 因而具有优异的耐热、耐氧、耐臭氧和耐天候老化性能。

[3] Heat aging resistance up to 130℃ can be obtained with properly selected sulfur acceleration systems and heat resistance at 160℃ can be obtained with peroxide cured compounds. 选择适当的硫黄促进剂体系, 得到的乙丙胶耐热温度可以高达 130℃, 而用过氧化物硫化则可耐 160℃。

Exercises

1. Read the following in English

1960's	60℃	-150℃	5-200+	ML 1+4 @ 125℃
45 %	80 wt. %	0.855	MPa	g/ml
>	<	>>	kg/m^3	

2. Put the following words or expressions into English

轮胎	胶管	纯胶管	胶带
胶鞋	橡胶工业制品	耐热老化	耐臭氧老化
耐天候老化	非晶橡胶	结晶橡胶	柔性
硫黄硫化体系	硫黄给予体硫化体系	过氧化物硫化体系	门尼黏度
硬度	拉伸强度	耐磨性	伸长率
永久变形	回弹性	分子量	分子量分布

3. Put the following words or expressions into Chinese

roofing membrane	plastic impact modification	thermoplastic vulcanizate
electrical resistivity	polar solvent	diene content
ethylene content	specific gravity	tear resistance
oil swell resistance	flame retardance	green strength
crystallinity	scorch resistance	collapse resistance
heat stability	die swell	modulus

4. Translation

EPDM (Ethylene Propylene Diene Monomer)

Originally developed in 1950's for tyre applications. Became more widely used because of its suitability for outdoor use.

Properties
- the most water resistant type of rubber - also very resistant to most water based chemicals
- very inert structure, remains stable over long periods of time
- can withstand temperatures of up to 130℃ for extended periods of time (months)
- very good weathering resistance
- easily compounded and processed

Limitations
- will not resist oil or oil based products
- compression set not as good as some other rubbers, but can be improved by careful compounding

Typical Applications
General engineering without exposure to oil

[Reading Material]

Surface Bloom of Rubber

Q1: A surface bloom has developed on stored rubber products. What is it likely to be and is it safe to handle?

A1: A true bloom occurs because a migratable substance is present in the rubber at a level above its solubility. This feature is exploited with hydrocarbon waxes in order to form a protective bloom against atmospheric ozone attack. If the bloom is removed by cleaning, ozone resistance will fall. Much less desirable is a bloom of antioxidant or vulcanization-reaction products, usually accelerator residues. Foremost among the latter are zinc dialkyldithiocarbamates and the zinc salt of mercaptobenzothiazole, since many of these are only sparingly soluble in some rubbers. A bloom of elemental sulphur may occur if the rubber compound has not been sufficiently vulcanized and so remains 'undercured'. Zinc stearate (a reaction product of zinc oxide and stearic acid, and sometimes used as a activator in its own right) may also migrate to the rubber surface if it is present in excessive levels.

Surface accumulation of a substance can also occur from chemically induced diffusion. A good example of this is the migration of a *p*-phenylenediamine type antiozonant to replenish that consumed at the surface by ozone attack.

Blooms and surface deposits, especially those of an unexpected nature, should always be treated with caution in view of concerns about toxicity or allergic response and risk of contamination. Some products will call for complete absence of a bloom.

Q2: Q1 considered the origin and composition of surface blooms. How can I ensure a completely bloom-free rubber product?

A2: The key is to prevent the surface migration of compounding ingredients and vulcanization-reaction products. Thus it will be necessary to avoid or certainly minimize the use of sparingly soluble additives, notably some vulcanization accelerators and some antidegradants. The same principle applies to accelerators and sulphur donors that give rise to sparingly soluble reaction products. It is also necessary to avoid materials, such as antiozonants, that are chemically drawn to the surface.

Two approaches can be adopted. One is to use additives that are more rubber-soluble, for example, a zinc dithiocarbamate having a longer alkyl chain, such as butyl or nonyl. The other is to use a blend of additives so that each additive can be kept at a level below its solubility limit. An example is a combination of two or more different dithiocarbamate accelerators. The synergism between some types of accelerators, for example, sulphenamides and thiuram sulphides, and between some types of antioxidants can also be exploited since this enables the overall additive levels to be reduced.

A more radical approach is to extract the product after vulcanization so that no substances are available for blooming or, indeed, for any type of migration. Some products, indeed, are solvent extracted to avoid risk of contamination or incompatibility. Obviously special measures, especially in choice of rubber type and vulcanizing system, may be needed to retain a measure of resistance to oxidative ageing.

Words and Expressions

bloom	[bluːm]	*n. / v.*	喷霜
handle	[ˈhændl]	*vt.*	运用, 处理, 操作
migrate	[maiˈgreit, ˈmaigreit]	*v.*	移动, 移往
exploit	[iksˈplɔit]	*v.*	使用
residue	[ˈrezidjuː]	*n.*	残余物
foremost	[ˈfɔːməust]	*adv.*	首要地, 首先
zinc dialkyldithiocarbamate			二烷基硫代甲酸锌

sparingly		adv.	部分地
undercure	[ˈʌndəkjuə]		欠硫
excessive level			过量
accumulation	[əkjuːmjuˈleiʃ(ə)n]	n.	积聚,堆积
deposit	[diˈpɔzit]	n.	堆积物,沉淀物
		vt.	存放,堆积
toxicity	[tɔkˈsisiti]	n.	毒性
allergic	[əˈləːdʒik]	adj.	过敏的
contamination	[kənˌtæmiˈneiʃen]	n.	污染
origin	[ˈɔridʒin]	n.	起源,由来,起因
composition	[kɔmpəˈziʃen]	n.	成分
notably	[ˈnəutbəli]	adv.	显著地,特别地
draw	[drɔː]	vt.	汲取,领取,吸引
approach	[əˈprəutʃ]	n.	方法,步骤,途径
nonyl	[ˈnɔnil, ˈnəu-]	n.	壬基
synergism	[ˈsinədʒizm]	n.	合作,并用
radical	[ˈrædikəl]	adj.	根本的,激进的
extract	[iksˈtrækt]	vt.	抽提

PART F   Polymer Processing

Unit 1   Extrusion

Extrusion is one of the most important plastics processing methods in use today. Most plastic material in the world are processed in extruders. [1]A substantial percentage of all plastics products base through two or more extruders on their way from the chemical reactor to the finished product.

Among the products manufactured in extruders are pipe, rod, film, sheet, fiber, and an unlimited number of shapes or profiles. However, the extruder is also used in compounding and in the production of plastic raw material such as pellets (also for reclaimation of scrap plastics).

[2]In addition to the continuous products mentioned above, for which the extruder is particularly suited, it is also used to manufacture noncontinuous products, either by using special molds, as in the case of blow-molding machines, or by reciprocating the screw, as in the case of reciprocating-screw injection molding.

Because the extruder is the basis of all these processing techniques, understanding its operation is extremely important.

The heart of the extruder is the Archimedean screw rotating in a barrel. It is capable of pumping a material under a set of operating conditions at a specific rate. After the pumping action and the shaping of the extrudate by dies the finished product is taken up by one of several take-up devices.

Computerization of the single-screw extruder

It can be seen that, contrary to popular belief, the plasticating extruder is not merely a simple pump, but a complex system comprised of a slow moving solid in a hopper, a solids-conveying zone, delay in melting zone, melting zone, melt-pumping zone, and often also a vent zone. Every portion of the extruder contributes very significantly to the pumping and pressure raising capacity of the extruder. However, they also all contribute to plasticating, mixing, heat transfer, and viscous heat dissipation. The interaction of these portions is therefore so complex that no manual computation is really capable of describing what is going on in an extruder. A computerized physical model extruder had to be constructed, therefore, to show the intricate of the whole process and the effect of each of its components on the overall performance of an extruder. This computerized model of the extruder permits us to computer simulate the operating conditions, extruding any resin. The simulator is also applicable to blow molding and injection molding.

Twin-screw extruders

In certain special application such as rigid PVC, it was found that single screw extruders heat the plastic to higher temperature than it is capable of withstanding. However, certain twin-screw extruders do not overheat the plastic. Several types of twin-screw extruders are on the market.

A. Nonintermeshing screws

In this type of extruder the pressure versus output relationship is similar to that in single-screw extruders and resembles, therefore, the operating characteristics of a centrifugal pump.

B. Intermeshing co-rotating screws

In this type of extruder the screw profile is often conjugated, that is no gap is left between the screws. As a result tolerances between screw, as well as between barrel and screw, can be smaller, reducing the residence time of the plastic and reducing thermal degradation of particularly sensitive grades. The material does not pass between the screws, but as soon as it reaches the other screw, continues to travel down the channel with it.

C. Intermeshing, counter-rotating screws

Here the two screws penetrate each other, but their centers are slightly more than a screw diameter apart. Since the plastic material passes the clearance between the screws it becomes less of a positive-displacement pump and its pumping capacity depends somewhat on the pressure. In addition the existence of the clearance also prevents one screw from wiping the other clean.

Twin-screw extruders combine heat generation by viscous heat dissipation with the solid and probably using up this sensible heat to melt some of the remaining solid plastic. This they tend toward more uniform temperatures with greater control. Their disadvantage, however, is their considerably higher cost, compared to single screw extruders, and the lack of a thorough understanding of what is really occurring inside the extruder during processing.

The availability of a good theoretical model for the single-screw extruder enables us to locate mixing devices strategically and predict their effects. As a result, in many cases the design of screws using modern computerized methods can eliminate the need for expensive twin-screw extruders.

New words

extrusion	[eks'truːʒən]	n.	挤出
profile	['prəufail]	n.	剖面，侧面，外形，轮廓
pellet	['pelit]	n.	粒料
reclaimation	[ˌreklə'meiʃən]	n.	再生，回收
reciprocating	[ri'siprəkeitiŋ]	adj.	往复的
centrifugal	[sen'trifjugəl]	adj.	离心的
screw	[skruː]	n.	螺杆
die	[dai]	n.	口模，口型
intricate	['intrikit]	adj.	复杂的，错综的
viscous	['viskəs]	adj.	黏性的，胶黏的
intermesh	[ˌintə(ː)'meʃ]	v.	使互相啮合
conjugated	['kɔndʒugeitid]	adj.	共轭的，成对的
gap	[gæp]	n.	缺口，间隙
penetrate	['penitreit]	v.	渗透，浸染
dissipation	[ˌdisi'peiʃən]	n.	消散，分散
uniform	['juːnifɔːm]	adj.	均匀的，一致的
strategically	[strə'tiːdʒikəli]	adv.	战略上

Notes

[1] A substantial percentage of all plastics products base through two or more extruders on their way from the chemical reactor to the finished product. 大多数的塑料制品从原料到制成最后产品都要经过两次或多次挤出过程。

[2] In addition to the continuous products mentioned above, for which the extruder is particularly suited, it is also used to manufacture noncontinuous products, either by using special molds, as in the case of blow-molding machines, or by reciprocating the screw, as in the case of reciprocating-screw injection molding. 挤出机除了特别适合上述产品的连续生产，也可配以专门模具（如在吹塑机中）或使用往复式螺杆（如往复螺杆式注塑中）生产非连续产品。

Exercises

1. Translate the following passage into Chinese

Both injection and blow molding typically rely on longer part runs to justify tooling cost and setup time, but nowadays long part runs are not necessarily attractive. Product-manufacturing companies have learned the merits of reduced inventories and just-in-time manufacturing. Today, sheet-fed thermoforming offers comparable sophistication in finished parts, and has the advantages of lower machinery and tooling costs and faster setup. Thus it can be an attractive alternative.

2.Write an abstract in English for the text

Reading Material

Extrusion

Extrusion

For the extrusion process, plastic in the form of pellets or powder are fed into a heated barrel where rotating screws homogenize it and squeeze it through a die to give a finished or semi-finished product. In most of cases, pellets are used in single screw extruders, and powders are used in twin screw extruders. The die is designed to produce the desired shape of the end product.

The extrusion process may be very easily visualized perhaps the best examples are the meat grinder and the toothpaste tube. However the very simplicity of these examples belies the complexity of the process and extrusion technology is highly developed both as a science and as an art.

The extrusion line

The extrusion line begins with the hopper which holds the plastic material (in either powder or granule form). The hopper continuously feeds the material to a heated barrel which contains a rotating screw. This screw transports the polymer to the die head and simultaneously the material is heated, mixed. Needless to say the detailed design of an extruder screw is extremely complex in order to perform all the above tasks. At the die the polymer takes up the approximate shape of the article and is then cooled either by water or air to give the final shape. As the polymer cools it is drawn along by haul-off devices and either coiled (for soft products) or cut to length (for hard products).

Extrusion processes

The typical extrusion line described above can be used, with modifications, to produce a wide variety of products. Some typical examples are described below.

Wire coating: for all types of wires and cables

Monofilament: for rope, bristles and synthetic textile fibres.

Blown film: for plastic bags, plastic film and heat shrinkable film for food packaging.

Sheet extrusion: for sign production, refrigerator interiors and even small boat hulls. When a clear sheet is produced it can be used in glazing or lighting applications.

Pipe and tube: plastic tubing is used for garden hose, industrial hose, food and drink, transport and hydraulic or pneumatic control. Plastics pipe is used for water, gas, agricultural drainage, sewers and drains. New developments allow plastics pipe to replace copper pipe for heating and hot and cold water services.

Materials

Most common thermoplastic polymers can be used for extrusion and the material choice is dependent on both the performance requirements and on the economic constraints. The most commonly used material for general-purpose extrusions is PVC. The wide application of this material is due to cost, chemical resistance and its availability in various hardnesses and colours.

The hardness of PVC can vary from the rigid type used for windows (Shore 'A' hardness of 100) to the plasticized or soft version used for garden hoses (generally Shore 'A' 80 degree) and even down to very soft materials of Shore 'A' 60 degree which have limited uses. The colour can be either matched to a colour sample or chosen from several hundred standard colours. PVC is a very versatile material but, as with all materials, there are limitations.

Words and Expressions

homogenize	[həˈmɔdʒənaiz]	vt.	均化
semi-finished product			半成品
visualize	[ˈvizjuəlaiz, ˈviʒ-]	vt.	想象
meat grinder	[ˈgraində]	n.	绞肉机
belie	[biˈlai]	v.	掩饰
simultaneously	[siməlˈteiniəsly]	adv.	同时地
plasticized	[ˈplæstisaiz]	v.	增塑的
monofilament	[ˈmɔnəuˈfiləmənt]	n.	单(根长)丝, 单纤(维)丝
bristle	[ˈbrisl]	n.	短粗纤维
shrinkable	[ˈʃriŋkəbl]	adj.	会收缩的
glazing	[ˈgleiziŋ]	n.	玻璃装配业, 玻璃窗
pneumatic	[nju(ː)ˈmætik]	adj	充气的

Unit 2 Molding

Mo(u)lding consists of confining a material in the fluid state to a mold where it solidifies, taking the shape of the mold cavity.

Injection molding is the most common means of fabricating thermoplastic articles. The molding compound, usually in the form of pellets that are approximately 1/8in. cubes or cylinders 1/8-in. in diameter by 1/8-in. in length (molding "powder"), is fed from a hopper (which may be heated with circulating hot air to dry the material) to an electrically heated barrel. The pellets are conveyed forward through the barrel by a rotating screw. The material is melted as it goes by a combination of heat from the barrel and the shearing (viscous energy dissipation) of the screw. Older machines had a reciprocating plunger in place of the screw. All the heat for melting had to be supplied by conduction from the barrel, and mixing was very poor. This resulted in low plasticizing (melting) rates and a thermally nonuniform melt. The screw generates heat within the material and provides some mixing, thereby increasing plasticizing rates and giving a much more uniform melt temperature. [1]Molten material passes through a check valve at the front of the screw, and as it is deposited ahead of the screw, it pushes the screw backward while material from the previous shot is cooling in the mold. The cooled parts are ejected from the mold, the mold closes, and the screw is pushed forward hydraulically, injecting a new shot into the mold.

The molten polymer flows through a nozzle into the water-cooled mold, where it travels in turn through a sprue, runners, and a narrow gate into the cavity. When the part has cooled sufficiently, the mold opens, the knockout pins eject the parts, with sprues and runners attached. The sprues and runners are usually removed by hand and, with thermoplastics, chopped and recycled back to the hopper as regrind. Depending on the extent of material degradation in the cycle and the property requirements of the finished part, injection molding operations may tolerate up to 25% regrind in the hopper feed.

The molds are opened and closed by hydraulic cylinders, toggle mechanisms, or combinations of both. These have to be pretty hefty because pressures in the mold can reach many thousand pounds per square inch. The molds themselves often involve much intricate hand labor and thus can be quite expensive, but when amortized over a production run of many parts, the contribution of mold cost to the cost of the finished item may be insignificant.   批量生产

Injection molding presses are rated in terms of tons of mold-clamping capacity and ounces (of general-purpose polystyrene) of shot size. They range from 2ton, 0.25oz laboratory units to 3500ton, 1500oz monsters that are used to mold garbage cans, TV cabinets, furniture, and so forth. They are usually completely automated. Cycle times (and therefore production rates) sometimes depend on plasticizing capacity (the rate at which the material can be melted), but more often than not they are limited by the cooling time in the mold, which is in turn established by the thickness of the part and the (usually very low) thermal diffusivity of the material. Typical cycle times are between 1 min and 2 min. As usual, there are compromises involved.[2]It is often tempting to try to reduce cooling time by lowering the mold temperature, thereby increasing the cooling rate, and by lowering the temperature at which the material is injected into the mold, reducing the amount it must be cooled before solidifying enough to allow removal from the mold without distortion. The higher material viscosity that results, however, can give rise to short shots (incomplete mold filling), poor surface finish, and part distortion from frozen-in strains.   锁模力/通用聚苯乙烯 注射量/实验用注射机 大型注射机/垃圾桶/电视机外壳 注塑成型周期 塑化能力 热扩散系数 短射（欠注）/表面光洁度 冻结应变

New Words

molding	['məuldiŋ]	n.	模塑，模压，成型
solidify	[sə'lidifai]	v.	(使)凝固，(使)固化
cavity	['kæviti]	n.	模腔
cylinder	['silində]	n.	机筒
hopper	['hɔpə]	n.	料斗
plunger	['plʌndʒə]	n.	柱塞
hydraulically	[hai'drɔ:likəli]	adv.	液压
sprue	[spru:]	n.	主流道
valve	[vælv]	n.	阀
nozzle	['nɔzl]	n.	喷嘴
diffusivity	[difju'siviti]	n.	扩散能力
runner	['rʌnə(r)]	n.	分流道
distortion	[dis'tɔ:ʃən]	n.	扭曲，变形
knockout pin		n.	顶销，顶杆
hefty	['hefti]	adj.	有力的，很重的
amortize	[ə'mɔ:taiz; æ'mərtaiz]	v.	分期清偿

Notes

[1] Molten material passes through a check vale at the front of the screw, and as it is deposited ahead of the screw, it pushes the screw backward while material from the previous shot is cooling in the mold. 熔融的物料流经螺杆前端的止回阀，并且当物料在螺杆顶端积存到一定量就把螺杆顶回去，同时先前注射的料在模具中冷却。

[2] It is often tempting to try to reduce cooling time by lowering the mold temperature, thereby increasing the cooling rate, and by lowering the temperature at which the material is injected into the mold, reducing the amount it must be cooled before solidifying enough to allow removal from the mold without distortion. 人们常试图通过降低模具温度提高冷却速度，和通过降低原料注射进入模具的温度减小冷却量的方法来缩短冷却时间，从而保证制品从模具中取出后充分固化而不会变形。

Exercises

1. Translate the following passage into Chinese

Injection molding machines are increasingly being equipped with feedback control systems that monitor cavity pressure, screw position, and screw velocity and control these quantities through the hydraulic system to provide high-quality, uniform parts, using a minimun amount of material and minimizing the number of rejects.

2. Please write out at least 3 kinds of plastics processing both in English and in Chinese

3. Give a definition for each following word

 (1) injection molding (2) extrusion

[Reading Material]

Rotational, Fluidized-Bed and Slush Molding

Molding techniques have been developed to take advantage of the availability of finely powdered plastics, mainly polyethylene and Nylons. In rotational molding, a charge of power is introduced to a heated mold, which is then rotated about two mutually perpendicular axes. This distributes the powder over the inner mold surfaces, where it fuses. The mold is then cooled by compressed air or water sprays, opened, and the part is ejected. Rotational molding is capable of producing extremely irregular hollow objects. The molds are inexpensive, often simply sheet metal, because no elevated pressures are involved, and they can be heated in simple hot-air ovens. Thus capital outlay is relatively low. The process is in many ways competitive with extrusion blow molding for the production of large hollow items such as drums and gasoline tanks. Blow molding requires a much large initial investment but is capable of higher production rates.

These two processes also provide a good example of how the processing operation, polymer, and finished part properties are often intimately connected. Blow molding can handle very high-molecular-weight linear polyethylenes, in fact often requires high-molecular-weight material to prevent excessive parison sag. Rotational molding, on the other hand, requires a low-molecular-weight resin because a low viscosity is needed to permit fusion of the powder under the influence of the low forces in a rotational mold. When subjected to stresses for long periods of time, particularly in the presence of certain liquids linear polyethylene has a tendency to fail through stress cracking. It turns out that high-molecular-weight resins are much more resistant to stress cracking. Thus, although the parts might appear similar, those produced by extrusion blow molding will ordinarily have superior resistance to stress cracking. The difference can be narrowed or eliminated by using more material (thereby lowering stress for a given load) or by using a crosslinkable polyethylene in rotational molding. Either of these solutions increases material cost, however.

Polymer powders are also used in a process known as fluidized-bed coating. When a gas is passed up through a bed of particles, the bed expands and behaves much like a boiling liquid. When a heated object is dipped into a bed of fluidized particles, those that contact it fuse and coat its surface. Such 100%-solids coating processes are increasing in importance because they eliminate the pollution often caused by solvent evaporation when ordinary paints are used.

Similar procedures have been in use for years with plastisols. Liquid plastisol is poured into a heated female mold. The plastisol in contact with the mold surface solidifies, and the remainder is poured out for reuse. This slush molding process is used to produce objects such as doll's heads. Dipping a heated object into a liquid plastically coats it with plasticized polymer. Vinyl coated wire, dish racks are familiar products of this process.

Words and Expressions

rotational molding			旋转模塑
fluidized-bed molding			流化床模塑
slush molding			搪塑
nylon	['nailən]	n.	尼龙
oven	['ʌvən]	n.	烤箱, 烤炉, 灶
perpendicular	[ˌpə:pən'dikjulə]	adj.	垂直的, 正交的
irregular	[i'regjulə]	adj.	不规则的, 无规律的
outlay	['autlei]	n.	费用
parison sag			型坯下垂
boiling	['bɔiliŋ]	adj.	沸腾的, 激昂的
dip	[dip]	v.	浸, 蘸, 沾
solvent	['sɔlvənt]	adj.	溶解的, 有溶解力
evaporation	[iˌvæpə'reiʃən]	n.	蒸发(作用)
coat	[kəut]	vt.	涂上, 包上

Unit 3　Rubber Processing

橡胶加工

1. Raw material weighing　　　　　　　　　　　　　　　原材料称量

[1]Raw rubber and compounding ingredients are weighted up to appropriate formulation to produce engineering rubber materials with specific properties.

2. Compounds mixing　　　　　　　　　　　　　　　　混炼

Mixing of the raw rubber and compounding ingredients is carried out on a <u>mill</u> or in an <u>internal mixer</u> to produce <u>rubber compounds</u>.　开炼机/密炼机/胶料

3. Preforming　　　　　　　　　　　　　　　　　　　预成型

The rubber compound is <u>sheeted</u> and cut or preformed to produce 下片
<u>blanks</u> suitable for moulding into finished parts. At this stage the rubber 胶坯
compound is unvulcanized and when warm is easily squeezed into shape.

4. Moulding　　　　　　　　　　　　　　　　　　　　模塑成型

There are three main methods of moulding:

　Extrusion moulding　　　　　　　　　　　　　　　挤出成型

Rubber compound is heated up in heating <u>cylinder</u>, fed forward 机筒
continuously by <u>screw</u>. [2]The moulded product in the shape of regular 螺杆
<u>sectional form</u> is pressed out through <u>die</u> by the rotation of the screw 断面形状/口型
and inner pressure.

　Injection moulding　　　　　　　　　　　　　　　注射成型

Heated compounding material is injected under pressure through a series of feed gates into the cavity itself.

　Compression moulding　　　　　　　　　　　　　模压成型

In which a rubber blank is placed between two <u>halves</u> of a mould 半模
and squeezed into shape under heated up and compressed by the <u>press</u>. 平板硫化机

5. Finishing　　　　　　　　　　　　　　　　　　　修整

[3]Rubber <u>mouldings</u> normally require a finishing operation to 模型制品
remove excess rubber or "<u>FLASH</u>" resulting from <u>tool split lines</u> or <u>feed</u> 飞边/模具分型面
<u>gates</u>. A variety of methods are used which fall into three main 注胶口
groupings:

　<u>Deflashing</u> in which parts are frozen at temperature below -80℃ 修边
and tumbled in their brittle state to remove excess rubber.

　<u>Buffing</u> using abrasive techniques such as <u>mops</u> and abrasive belts 抛光/抛光轮
or wheels.

　<u>Cutting</u> which involves a wide range of methods from 冲裁
hand-scissors and punching to knife-form in <u>clicking presses</u>, 冲裁机
semi-automated knife systems, etc.

6. Inspecting and testing

Rubber parts are carefully examined and checking :

- Size and dimension
- <u>Tensile Strength</u>, <u>tensile set</u>
- <u>Modulus</u>
- Density, <u>specific gravity</u>
- Chemical, <u>fluid and weather resistance</u>
- Hardness
- <u>Elongation</u>
- <u>Compression set</u>
- <u>Resilience</u>

检测
拉伸强度/拉伸永久变形

定伸应力/压缩永久变形

密度/回弹性
耐介质性/耐候性

New Words

preforming	[pri'fɔːmiŋ]		预成型
sheet	[ʃiːt]	vt.	压片
blank	[blæŋk]	n.	胶坯
moulding	['məuldiŋ]	n.	成型
cylinder	['silində]	n.	机筒
feed	[fiːd]	vi.	进料，喂料，流入
press	[pres]	n.	平板硫化机
finishing	['finiʃiŋ]	n.	修饰，整理
split line	[split]		分型线
fall into			分成，属于
deflashing			修边，除边
tumble	['tʌmbl]	vt.	使滚翻，弄乱
buffing	['bʌfiŋ]		磨削，抛光
abrasive	[ə'breisiv]	n.	研磨剂
mop	[mɔp]	n.	adj. 研磨的（抛光用）毛布轮
cutting	['kʌtiŋ]	n.	切割，冲裁
punching		n.	冲孔，冲眼

Notes

[1] Raw rubber and compounding ingredients are weighted up to appropriate formulation to produce engineering rubber materials with specific properties. 生胶和配合剂按照配方进行称量,然后制成具有特定性能的橡胶工程材料。

[2] The moulded product in the shape of regular sectional form is pressed out through die by the rotation of the screw and inner pressure. 具有一定断面形状的模型制品是胶料在螺杆的转动和压力作用下从口型中挤出得到的。

[3] Rubber mouldings normally require a finishing operation to remove excess rubber or "FLASH" resulting from tool split lines or feed gates. 橡胶模型制品通常要进行修整，除去由模具的分型线或流胶口造成的飞边。

Exercises

1. Translate the following into Chinese

Key processes in manufacture are storage, weighing of ingredients, mixing, shaping and vulcanization using processes such as moulding, extrusion and calendering.

Rubber may also be processed in the liquid latex concentrate form with ingredients added as dispersions, emulsions or solutions before the product is manufactured by dipping, casting, foaming or extrusion.

The last step is to vulcanize the rubber, which gives it strength, hardness and elasticity by treating it with heat and sulfur compounds. The more sulfur compound added, the firmer the vulcanized compound will be. Manufacturers vulcanize and shape molded products by heating the molds under pressure.

2. Translation

Mixing

Some rubbers may require a breakdown or mastication step prior to addition of other formulation components. The mastication involves a reduction in molecular weight and a reduction in viscosity to make the material more easily processable. Mixing temperature is dependent upon the type of elastomer used. First stage mixing temperatures in an internal mixer are often in the range 110~125℃.

Following a high temperature mixing of rubber, carbon black and oil, a second stage mixing is often done at a lower temperature to add sulphur and curatives. This second stage mix may be done in an internal mixer or on a mill. A rubber mill consists of two temperature controlled steel rollers. The high shear condition to provide good mixing and dispersion.

The lower temperature used in a second stage mixing prevents the onset of vulcanization that could otherwise take place in the presence of the curatives; exceptions to two-step mixing are cases where stocks have a low scorch tendency or where a properly chosen pre-vulcanization inhibitor is used.

A homogeneous dispersion of all components is essential to achieve consistency throughout the final product. Incomplete dispersion will lead to localized concentrations of ingredients which can lead to physical imperfection, poor tensile, tear, heat build-up, etc.

3. Translation

Rubber Processing

Some operations employed in the manufacture of rubber articles are completely analogous to those in plastics technology. This is because in conventional rubbers the rubber compound or mix is predominantly plastic before vulcanization starts. Vulcanization converts the rubber from the predominantly plastic state to the predominantly elastic state. In most rubber applications this transformation is necessary to obtain the desired and durable rubber properties, and time must be allowed for this transformation which is called curing or vulcanization. Thus, in some respects conventional rubbers resemble the thermosetting plastics. This means that rubber processing is slower than the processing of thermoplastics. For this reason, modern developments in synthetic rubbers have strive towards thermoplastic elastomers which need no chemical curing systems or towards self-curing liquid rubbers.

[Reading Material]

Processing Rubber

1. Compounding

Raw rubbers have few uses in their natural state. To achieve the desired range of properties, the raw rubber must be combined with a range of additives. The selection of appropriate additives, and their skilful and consistent mixing, is known as compounding.

2. Mixing

The constituents are weighed out and combined by a mixing process which must blend the ingredients thoroughly in a repeatable way. This is achieved either by an internal mixer, where the compound is mixed by two meshing rotors in an enclosed case; or by open mill mixing, adding the ingredients carefully into the "nip" between two steel rollers, typically of 30in diameter.

The result of either process is a batch of uncured rubber compound. This is allowed to settle for a time before undergoing Quality Assurance tests. Once passed, it can be formed into suitable shapes for moulding.

3. Pre-Forming

Each moulding process has its own requirements for uncured material. Compression moulding, for example, requires a "blank" of material in a size which will fill the cavity exactly. Direct injection moulding needs relatively large quantities of compound in a continuous strip. Due to the nature of the injection process, material properties must be precisely measured and controlled to achieve the planned flow and cure behaviour, as well as the desired final characteristics of the rubber.

A variety of processes are used to produce material suitable for moulding:

Sheeting Uncured material is produced in sheets of the desired thickness. Sometimes "blanks" are cut from the sheet, like pastry cutting.

Extrusion Extruders force warmed compound through a shaped die. Any reasonable length of shaped material can be produced. Once cooled this is fed into the direct injection presses.

Pre-Forming Extrusions as above are cut to required lengths as they emerge from the die. This process can be accurately controlled to produce blanks of precise volume for compression moulding.

4. Moulding

Rubber can be made into components by a number of processes, including extrusion, calendering, coating onto fabric and moulding.

The choice of moulding method will depend upon factors such as the finish desired, the number of components required, the money available for tooling etc.

The basic processes of moulding are as below.

Compression A piece of uncured rubber of the correct size is placed between two halves of a heated mould. The mould is closed in a press under a pressure of around one ton/sq in and the rubber is forced into the exact shape of the cavity. The rubber gains heat by conduction from the mould surfaces and "cured". When the rubber had had sufficient time to cure, the mould can be opened and the part removed.

Compression moulding is a relatively simple process and is often used for components required in fairly low quantities. It is also the most economic method for parts with simple shapes. Parts moulded by this method will always have some flash because the mould surfaces are held apart by the necessary excess rubber in the "bank".

Transfer injection The heated mould is closed in a press and the rubber injected by a hydraulic cylinder through a feed hole in the cavity. The cylinder can either be incorporated in the press or sometimes in the mould. Provision must be made for air to escape from the cavity and the rubber enters, and the feeding method chosen to suit the operational requirements of the part.

This method of moulding can produce high-precision parts in moderate quantities without high tooling costs. In the simplest case, the mould can be the same as a compression mould with the addition of a feed hole. Maximum weights and number of cavities are governed by the capacity of the transfer cylinder and the clamp pressure.

Direct injection A screw injection system delivers a metered quantity of rubber into the closed mould. The injection unit is fed from a continuous strip or a reservoir of uncured rubber and is cooled to avoid premature curing.

This process is generally used for multi-cavity moulds and can produce hundreds of components per press cycle. Because of the amount of rubber in the system, it is inadvisable to change materials frequently.

Large moulds require complex feed systems to balance the pressures in each cavity. Generally these are in the heated top half of the mould and cure at the same time as the components. Unlike thermoplastic, cured thermoset rubber cannot be reground or reused and the additional waste has to be included in the material usage per piece. Where very large volumes of mouldings are required, cold runners system should be considered.

This process lends itself to relatively large quantities, a large number of cavities and infrequent change of materials or moulds. Parts are repeatable and can be made to a high level of precision.

5. Finishing

Compression moulded and some injection moulded parts require deflashing. This is done in various ways depending upon the shape and size of the component and the type of rubber used.

Sub-Zero Finishing The most modern and efficient method of finishing uses cryogenics. Parts are frozen to temperatures as low as $-120\,^\circ\text{C}$ and then tumbled and/or bead-blasted while cold to remove the brittle flash. The machines are individually programmed with the optimum temperature and running times for each particular type and number of components, which are tested and proven during the development stages.

Tear Finishing The tool is designed to produce a very thin section of flash around the part which is torn off at the press during demoulding.

If required, components can also be hand finished by the following methods:

Buffing Parts can be smoothed using a variety of abrasive belts, wheels and mops.

Cutting Parts are die cut by machine using a range of tools from precision steel knives to wood-formes, or in extreme cases are trimmed with scissors.

Words and Expressions

constituent	[kən'stitjuənt]	n.	成分
mesh	[meʃ]	vt.	啮合
rotor	['rəutə]	n.	转子
nip	[nip]	n.	辊距
diameter	[dai'æmitə]	n.	直径
batch	[bætʃ]	n.	一批
quality assurance			质量保证
strip	[strip]	n.	条，带
pastry	['peistri]	n.	面粉糕饼，馅饼皮
emerge from			自…出现
incorporated	[in'kɔːpəreitid]		合成一体的
clamp	[klæp]	n.	夹子，夹具，夹钳
cryogenics	[kraiəu'dʒeniks]	n.	低温

Appendix Ⅰ The Nomenclature of Synthetic Elastomers

The classification system used in the rubber industry for naming elastomers is based on that described in ISO 1629—1976. The last letter of the identification code.(标识码) defines the basic group to which the polymer belongs whilst the earlier ones provide more specific information and in many cases define the polymer absolutely. The list below is not exhaustive(无遗漏的，详尽的) but covers many of the more common elastomers or rubbers and some "rubber-like" materials.

'M' Group: Rubbers with a Saturated—C—C—Main Chain:
IM:　　　Polyisobutylene , a soft inert polymer.
EPM:　　Copolymer of ethylene and propylene, rubber-like material.
EPDM:　　A terpolymer of ethylene, propylene and a di- or polyene giving pendent olefin groups as crosslinking sites.
CSM:　　Chlorosulphonated(氯磺化) polyethylene, containing both C—Cl and C—SO_2Cl groups. Cl content 20%～45%; S content 0.5%～2.5%. Optimum properties 30% Cl, 1.5% S; ozone-resistant rubber also used in varnishes(清漆).
FPM:　　Fluoro/fluoroalkyl groups on C—C backbone (e.g. copolymers of hexafluoropropylene(四氟丙烯)and vinylidene fluoride(偏氟乙烯); copolymer of vinylidene fluoride and 1-hydropentafluoropropylene (1-氢五氟丙烯).
CFM:　　As FPM, but containing Cl as well as F; vinylidene fluoride (VF): chlorotrifluoroethylene (CTFE，三氟氯乙烯) copolymer.

'O' Group: Rubbers with Carbon and Oxygen in the Main Chain:
CO:　　　(氯醚橡胶)Poly(epichlorohydrin)-the parent material from which comes.
ECO:　　Copolymer of epichlorohydrin (表氯醇)and ethylene oxide(环氧乙烷).
GPO:　　Copolymer of propylene oxide(环氧丙烷) and allyl glycidyl ether(烯丙基缩水甘油醚) .

These materials have good heat resistance and good low temperature properties.
'Q' Group: Silicone Rubbers:
MQ:　　　Polydimethylsiloxane(聚二甲基硅烷); depending on the molar mass this can be an oil, wax or rubber.
MPQ:　　as MQ with the addition of phenylmethylsiloxane(苯基甲基硅烷).
MPVQ:　　as above but with vinyl groups.

MFQ: as MQ but fluorinated(氟化).

These are all relatively stable thermally and because of their cold cure characteristics may be used as electrical insulants, seals, moulds, etc.

'R' Group: Rubbers with an Unsaturated Carbon Backbone:

ABR: Refers to copolymers of butadiene and methyl methacrylate used to impregnate paper but also includes the terpolymer with acrylonitrile (primer, before adhesive layer applied) and tetrapolymer(四元共聚物) with styrene (used as a synthetic rubber).

BR: Poly(butadiene) - available as high *cis* (98%+), high *trans* (98% +) and anywhere in between. Can also have vinyl groups present at any level. General-purpose rubbers usually 90%+ *cis* or about 45% *cis* 45% *trans* 10% vinyl. High vinyls have some specialist uses.

CR: Poly(β-chlorobutadiene). Two main types, 'G', amber in colour with large molar mass range centred at about 100 000; 'W', white, molar mass of narrower range and centred about 200000. Used as an adhesive or where oil or ozone resistance required; gaskets, subaqua(水下的) suits, etc.

IIR: Copolymer of isobutylene and diene such as butadiene or isoprene (Butyl). Only a small amount of diene added (ca 2%～5%) to give crosslinkable sites. Has low gas permeability, hence uses in inflatable products, and as general-purpose rubber.

CIIR: Chlorinated IIR   with 2%～3% w/w halogen to decrease gas.

BIIR: Brominated IIR   permeability and improve self-adhesion on building.

IR: Synthetic *cis*-poly(isoprene)
cis level 90%～99%, remainder *trans* and vinyl. General-purpose rubber.

NBR: Copolymer; acrylonitrile and butadiene available with a wide range of ACN loadings to alter hardness; oil-resistant applications. Also available is terpolymer and tetrapolymer with styrene.

NR: *Cis*-poly(isoprene) natural rubber, essentially 100% *cis*, *trans*/vinyl <0.1%. Contains about 95% polyisoprene. Various grades available RSS, SMR, SIR, SLR, NIG with number identifying grade—5, 10, 20. Also modified NR—PA, SP, OENR, ENR, DPNR. NR was the original general purpose (GP) rubber.

SBR: Random copolymer of styrene and butadiene. Styrene level varies from 10% to 80% but the general purpose level is 23.5%. Many types available and the exact type identified by a numeric code. General purpose rubber. Vast amounts used in tyres. Also available as ter/tetra polymer systems (see ABR & NBR).

'T' Group: Rubbers with Carbon, Oxygen and Sulphur in the Main Chain

OT: Polymer of *bis*-chloroalkylether(双氯烷基醚) (or formal 缩甲醛), with sulphur. Most common one uses *bis*-2-chloroethylformal; $CH_2(OCH_2CH_2Cl)_2$[with a little 1,2,3-trichloropropane(1,2,3-三氯丙烷) for crosslinking].

EOT: As above, but copolymerized with ethylene dichloride(1,2-二氯乙烷). All of these smell strongly of sulphur and are used for oil and solvent seals. The liquid polymers cold cure and are used as sealants in the building trade. Popular ones include Poly (ethylene disulphide)(聚二硫化乙烯) and Poly(butyl ether disulphide)

'U' Group: Polymer Chain Contains Carbon, Oxygen and Nitrogen

AU: Polyesterurethanes (聚酯型聚氨酯橡胶).

EU: Polyether urethanes (聚醚型聚氨酯橡胶).
A wide range of materials used as oil-resistant materials, in oxidation-resisting applications and as lightweight shoe soling.

Although not elastomers, certain other polymeric materials can exhibit "rubber-like" behaviour.

PVC: Poly(vinylchloride); hard brittle material ($d = 1.4$) often copolymerised with vinylidine chloride(偏氯乙烯), vinyl acetate(醋酸乙烯酯) styrene, ABR, ethylene vinyl acetate(乙烯乙酸乙烯酯) etc. for a wide range of applications. When plasticized, usually with esters such as phthalates(邻苯二甲酸酯), it becomes quite 'rubbery', used in conveyer belts, paints, varnishes, floor coverings, erasers (rubbers), flexible tubing, wellington boots(半长筒胶靴) and many cheap "rubber" goods. Thermoplastic.

PE: Polyethylene; a wide range of types available—HDPE (high density PE) and LDPE (low density PE). Numerous applications—medical implants to polythene bags, blended with elastomers such as EPDM to produce thermoplastic elastomers. Type distinguished by their melting points. LDPE <110℃, HDPE up to 136℃. A reclaimed mixture shows a spread of melting points.

PP: Polypropylene; similar applications to PE but higher melting (165℃). Also used to make thermoplastic elastomers.

PS: Polystyrene; occasionally met as a reinforcing plastic within a continuous elastomeric phase (e.g. shoe soling) but can be considered to be present in some thermoplastic elastomers such as the block copolymers: SIS styrene-isoprene-styrene SBS styrene-butadiene-styrene

Chlorinated rubber; refers specifically to chlorinated natural rubber. Used for paints and adhesives. The theoretical level for $(C_5H_8Cl_2)_n$ is 51% but commercial chlorinated rubber contains 65% Cl.

Rubber hydrochloride; again refers specifically to hydrochlorinated(氢氯化) natural rubber - usually with about 90% of the double bonds hydrochlorinated (30% Cl). Plasticized material produced as film was used for packaging.

M. G. Rubber; natural rubber to which methyl methacrylate has been grafted, commercial materials generally contain 30% or 49% w/w methacrylate.

GuttaPercha and Balata; (古塔波胶，杜仲树胶，巴拉塔胶) Polyisoprene, with 100% of the units *trans*; pure material not unlike PVC in feel and, when plasticized, can have similar uses.

Chicle; (糖胶树胶)a naturally occurring mixture of *cis* and *trans* polyisoprene (25 : 75), with resins, used in chewing-gum(口香糖).

Guayule; (银菊胶)natural *cis*-polyisoprene isolated from the shrub(灌木) *Parthenium argentatum* (银色橡胶菊)by solvent extraction(溶剂抽提). Uses and properties as for NR, but smell reminiscent(忆起) of gin(杜松子酒). Efforts to develop commercial exploitation have not been particularly successful.

Appendix II Acronyms of Polymers

ABS	Acrylonitrile-butadiene-styrene.
ACS	Acrylonitrile cholorinate polyethylene and styrene.
ASA	Acrylic-styrene-acrylonitrile.
CA	Cellulose acetate.
CAB	Cellulose acetate-butyrate.
CAP	Cellulose acetate-proplonate.
CPE	Chlorinated poly ethylene.
CPI	Condensation-reaction polyimides.
CPVC	Chlorinated polyvinyl chloride.
CTFE	Chlorotrifluoro ethylene.
EC	Ethyl cellulose.
ECTFE	Ethylene-chlorotrifluoro ethylene.
EMA	Ethylene-methyl acrylate.
EP	Ethylene propylene.
ESCR	Environmental stress crack resistance.
ETFE	Ethylene-tetrafluoro ethylene.
EVA	Ethylene-vinyl acetate.
FDA	Food & Drug Administration.
FEP	Fluorinated ethylene-propylene.
FR	Fiber reinforced.
FRP	Fiber-reinforced plastics.
HDPE	High-density polyethylene.
HIPS	High-impact polystyrene.
HM	High-modulus.
HMC	High-strength molding compound.
HME	High-vinyl modified epoxy.
HMW	High molecular weight.
IPN	Interpenetrating polymer network.
LCP	Liquid crystal polymer.
LDPE	Low density polyethylene.
LIM	Liquid injection molding.
LLDPE	Linear low-density polyethylene.
LMC	Low-pressure molding compound.
LMW	Low molecular weight.

Appendix II Acronyms of Polymers

MA	Maleic anhydride.
MBS	Methacrylate-butadiene-styrene.
MDPE	Medium-density polyethylene.
MMA	Methyl methacrylate monomer.
MW	Molecular weight.
PA	Polyamide (nylon).
PAI	Polyamide-imide.
PAN	Polyacrylonitrile.
PB	Polybutylene.
PBT	Polybutylene terephthalate.
PBTP	Polybutylene terephthalate.
PC	Polycarbonate.
PCTFE	Polychlorotrifluoroethylene.
PE	Polyethylene.
PEC	Polyphenylene ether copolymer.
PEEK	Polyetherether ketone.
PEH	Polyphenylene ether homopolymer.
PEI	Polyetherimide.
PEO	Polyethylene oxide.
PES	Polyethersulphone.
PET	Polyethylene terephthalate.
PETP	Polyethylene terephthalate.
PF	Phenyl-formaldehyde.
PFA	Perfluoroalkoxy (resin).
PI	Polyimide.
PIB	Polyisobutylene.
PMMA	Polymethyl methacryiate.
PMS	Paramethyldtyrene.
PMT	Polymethylpentene.
PP	Polypropylene.
PPO	Polyphenylene oxide.
PPS	Polyphenylene sulfide.
PS	Polystyrene.
PTFE	Polytetrafluoroethylene.
PU	Polyurethane.
PVC	Polyvinyl chloride.
PVDF	Polyvinylidene fluoride.
PVF	Polyvinyl fluoride.
RH	Relative humidity.
RH	Rockwell hardness.

RIM	Reaction injection molding.
RP	Reinforced plastics.
RTM	Resin-tansfer molding.
SAN	Styrene-acrylonitrile.
SMA	Styrene maleic anhydride.
SMC	Sheet molding compound.
TFE	Polytetra fluoroethylene.
TMC	Thick molding compound.
TPE	Thermoplastic elastomer.
TPU	Thermoplastic polyurethane.
UF	Urea-formaldehyde.
UHM	Ultra-high-modulus.
UHMW	Ultra-high molecular weight.
UV	Ultraviolet.
VAE	Vinyl acetate-ethylene.

Appendix Ⅲ Glossary of Polymer Engineering

Abrasion Resistance (耐磨性)—The ability of a material to withstand mechanical actions such as rubbing, scraping or erosion, that tend progressively to remove material from its surface.

Additive (添加剂)—In the plastics industry: materials added in minor amounts to basic resins or compounds to alter properties. A substance compounded into a resin to enhance or improve certain characteristics.

Alloy (合金)—A term used in the plastics industry to denote blends of polymers or copolymers with other polymers or elastomers.—i.e. ABS/Polycarbonate.

Ambient Temperature (环境温度) —The temperature surrounding an object and is often used to denote prevailing room temperature.

Amorphous (非晶的，无定形的)—Devoid of crystallinity or stratification. Most plastics are amorphous at processing temperatures.

Antioxidant (抗氧剂)—Additives which inhibit oxidation at normal or elevated temperatures.

Antioxidants & Antiozonants (抗氧剂和抗臭氧剂)—These additives are used to prevent the negative effects of oxygen and ozone on the resin materials.

Application (应用)—The act of applying or putting to use. What the molded plastic article will be in its final form.

Ash Content (灰分)—The solid residue remaining after a substance has been incinerated or heated to a temperature sufficient to drive off all combustible or volatile substances.

A-stage (甲阶，可熔阶段)—This is a very early stage in the reaction of certain thermosetting resins where the molecular weight is low and the resin is still soluble in some liquids and still fusible.

Biocides & Fungicides (抗微生物剂和防霉剂)—These additives act as pesticides and are used to inhibit the growth of fungus and other pests.

Blister (起泡)—An imperfection on the surface of a plastic article caused by a pocket of air or gas beneath the surface.

Bloom (喷霜)—An undesirable cloudy effect or whitish powdery deposit on the surface of a plastic article caused by the exudation of a compounding ingredient such as a lubricant, stabilizer pigment, plasticizer, etc.

Blow Molding (吹塑)—The process of forming hollow articles by expanding a hot plastic element called a parison (hollow tube) against the internal surfaces of a mold.

Blow Molding Method (吹塑成型)—fabrication in which a warm plastic parison (hollow tube), is placed between the two halves of a mold cavity and forced to assume the shape of that mold cavity by use of air pressure.

Blowing & Foaming Agents (发泡剂)—Upon addition to plastics or rubbers and then heating, this chemical generates inert gases which results in the resin assuming a cellular structure.

Branching (支化)—The growth of a new polymer chain from an active site on an established chain, in a direction different from that of the original chain.

Brittle Temperature (脆化温度)—The temperature at which materials rupture by impact under specified conditions.

B-stage (乙阶)—This describes an intermediate stage of reaction where the material will soften when heated and swells in the presence of certain liquids, but may not completely fuse or dissolve. The resin is usually supplied in this uncured state.

Bulk Density (堆积密度)—The density of a mod-king material in loose form expressed as a ratio of weight to volume.

Calendering (压延)—A form of extrusion using two or more counter rotating rolls in which film and sheet is produced by squeezing a hot, viscous material between them.

Casting (铸塑，浇注)—The process of forming solid or hollow articles from fluid plastic mixtures or resins by pouring or injecting the fluid into a mold or against a substrate with little or no pressure, followed by solidification and removal of the formed object.

Cavity (模腔)—A depression or a set of matching depressions, in a plastics-forming mold which forms the outer surfaces of the molded articles.

Charge (加料量)—The amount of material used to load a mold at one time or during one cycle.

Clamping Force (合模力)—In injection molding, the pressure which is applied to the mold to keep it closed, in opposition to the fluid pressure of the compressed molding material within the mold cavity and the runner system.

Clamping Pressure (锁模力)—In injection molding, the pressure applied to the mold to keep it closed during the molding cycle.

Co-extrusion (共挤)—The process of combining two or more layers of extrudate to produce a multiple layer product in a single step.

Cold Flow or Creep (冷流)—A time-dependent strain of solids resulting from stress.

Colorants (着色剂)—Dyes or pigments which impart color to plastics.

Colorants & Pigments (颜料)—Additives are used to change the color of plastic. They can be a powder or a resin/color premix.

Composite (复合材料)—An article or substance containing or made up of two or more different substances. Typically, one of the materials is a strengthening agent, the other being a thermoset or thermoplastic resin.

Compound (混合料)—These are chemical combinations of materials which include all the materials necessary for the finished product. They include BMC (Bulk Molding Compounds), SMC (Sheet Molding Compounds) and TMC (Thick Molding Compounds).

Compounding (混合)—The process required to mix the polymer with all of the materials that are necessary to provide the end user with a finished product.

Compression Molding (模塑)—The process of molding plastic in to a shape by applying pressure and heat. This process is most often used with thermoses.

Compressive Strength (压缩强度)—The ability of a material to resist a force that tends to crush it.

Continuous Service Temperature (连续使用温度)—The highest temperature at which a material can perform reliably in long term application - long term being, however, inconsistently defined by the manufacturers.

Copolymer (共聚物)—The chemical reaction of two different monomers combined each other, result in a compound.

Corrosion Resistance (耐腐蚀性)—A broad term applying to the ability of plastics to resist many environments.

Coupling Agents (偶联剂)—A material that is used to form a chemical bridge between the resin and glass or mineral fiber. By acting as an interface, bonding is enhanced.

Crazing (银纹)—Small cracks near or on the surface of plastic materials.

Creep (蠕变)—Due to its viscoelastic nature, a plastic subjected to a load for a period of time tends to deform more than it would from the same load released immediately after application, and the degree of this deformation is dependent of the load duration.

Cross-linking (交联)—The formation of chemical links between the molecular chains in polymers. This process can be achieved by chemical reaction, vulcanization, and electron bombardment.

Crystal (晶体)—A homogeneous solid having an orderly and repetitive three dimensional arrangement of its atoms.

Crystallinity (结晶度)—A state of molecular structure in some resins attributed to the existence of solid crystals with a definite geometric form, such structures are characterized by uniformity and compactness.

C-stage—This term describes the final stage of the reaction where the material s relatively insoluble and infusible.

Cure (硫化，固化)—The process of changing properties of polymer into a more stable and usable condition. This is accomplished by the use of heat, radiation, or reaction with chemical additives.

Cure Cycle (硫化周期，固化周期)—The time periods at defined conditions to which a reacting thermosetting material is processed to reach a desired property level.

Cycle Time (周期)—In a molding operation, cycle time is the time elapsing between a particular point in one cycle and the same point in the next cycle.

Degradation (降解)—A deleterious change in the chemical structure, physical properties or appearance of a plastic caused by exposure to heat, light, oxygen or weathering.

Density (密度)—The equivalent property to specific gravity; measured by displacement.

Dielectric Constant (介电常数)—The ratio of the capacity of a condenser made with a particular dielectric material to the capacity of the same condenser with air as the dielectric. Measured at a frequency of 10^6 cycles per second.

Dielectric Strength (介电强度)—A measure of the voltage required to puncture a material.

Durometer (硬度计)—An instrument used for measuring the hardness of a material.

Elasticity (弹性)—The ability of a material to quickly recover its original dimensions after removal of a load that has caused deformation.

Elastomer (弹性体)—A material which at room temperature can be stretched repeatedly to at least twice its original length and, upon immediate release of the stress, will return with force to its approximate original length.

Elongation, Break (断裂伸长)—The increase in distance between two gauge marks at the break point divided by the original distance between the marks. A zero value in the field indicates that it measured less than one.

Elongation, Yield (屈服伸长)—The increase in distance between two gauge marks at a yield point divided by the original distance between the marks. A zero value indicates that it measured less than one.

Engineering Plastics (工程塑料)—A broad term covering all plastics, with or wirhout fillers or reinforcements, which have mechanical, chemical and thermal properties suitable for use, in construction, machine components and chemical processing equipment.

Extender (增量剂)—A material added to a plastic compound used to reduce the amount of resin required per unit value.

Extrudate (挤出物)—The product or material delivered from an extruder, for example, film, pipe profiles.

Extruder (挤出机)—A machine for producing more or less continuous lengths of plastics sections such as rods, sheets, tubes, and profiles.

Extrusion (挤出)—The process of foaming continuous shapes by forcing a molten plastic material through a die.

Fabricating (二次加工)—The manufacture of plastic products by appropriate operations. This includes plastics formed into molded parts, rods, tubes, sheeting, extrusion and other forms by methods including punching, cutting, drilling, tapping, fastening or by using other mechanical devices.

Fatigue Strength (疲劳强度)—The maximum cyclic stress a material can withstand for a given number of cycles before failure occurs.

Filler (填料)—A relatively inert substance added to a plastic compound to reduce its cost and/or to improve physical properties, particularly hardness, stiffness and impact strength.

Fillers & Reinforcements (填料和增强剂)—Fillers are used to make a resin less costly. They can be inert or they can alter some properties of the plastic. Reinforcements are substances used to strengthen or give dimensional stability to a material.

Film (薄膜)—Films are flat materials that are extremely thin in comparison to its length and breadth. Typically, a film has a maximum nominal thickness of 0.25 millimeters.

Finish (光洁度)—The surface texture of a finished article.

Flame Retardants (阻燃剂)—Additives that reduce the tendency of plastics to burn.

Flash (飞边)—The thin, surplus of material which if forced into crevices between mating mold surfaces during a molding operation remains attached to the molded article.

Foaming Agent (发泡剂)—Any substance which alone or in combination with other substances is capable or producing a cellular structure in a plastic mass.

Forming (成型)—The process whereby a shape is transformed into another configuration.

Fracture (断裂)—The separation of a body, usually characterized as either brittle or ductile.

Gate (浇口)—In injection molding the channel through which the molten resin flows from the runner into the cavity.

Glass Fibers (玻璃纤维)—A family of reinforcing materials for reinforced plastics based on single filaments of glass.

Hardener (硬化剂)—A substance or mixture of substance added to a material to increase or control the curing reaction by taking part in it.

Hardness (硬度)—The resistance of a material to compression, indentation and scratching. There are several scales, and the data in the book gives both the scale used and the value on it.

Heat Stabilizers (热稳定剂)—These additives increase the ability of the material to withstand the negative effects of heat exposure. They are used to increase the overall service temperature of the material.

Impact Modifiers (抗冲改性剂)—Are additive used to enhance the material's ability to withstand the force of impact.

Impact Resistance (抗冲性)—The resistance of plastic articles to fracture under stresses applied at high speeds.

Impact Strength (冲击强度)—The ability of a material to withstand shock loading.

Injection Blow Molding (注吹)—Blow molding process by which the plastic parison to be blown is formed by injection molding.

Injection Molding (注塑)—The method of forming objects from granular or powdered plastics, most often of the thermoplastic type, in which the materials is fed from a hopper to a heated chamber in which it is softened, after which a ram or screw forces the material into a mold. Pressure is maintained until the mass has hardened sufficiently for removal from the mold.

Light, UV Stabilizers & Absorbers (光稳定剂，紫外光吸收剂)—These additives increase the ability of the material to withstand the negative effects of light and UV exposure, thus increasing the service life of the material.

Linear Thermal Expansion (线膨胀)—The fractional change in length of a material for a unit change in temperature.

Liquid Injection Molding (LIM) (液体注塑，反应注塑)—The process that involves an integrated system for proportioning, mixing, and dispensing two component liquid resin formulations and directly injecting the resultant mix into a mold which is clamped under pressure.

Low Temperature Flexibility (低温柔性)—The ability of a plastic to be bent without fracture at reduced temperatures.

Lubricant (润滑剂)—Internal lubricants, without affecting the fusion properties of a compound, promotes resin flow. External lubricants promote release from metals which aids in the smooth flow of melt over die surfaces.

Lubricant Bloom (润滑剂渗出)—Irregular, cloudy, greasy exudation on the surface of a plastic.

Master Batch (母料)—A concentration of a substance (an additive, pigment, filler, etc.) in a base polymer.

Mechanical Property (力学性能)—Properties of plastics which are classified as mechanical include abrasion resistance, creep, ductility, friction resistance, elasticity hardness, impact resistance, stiffness and strength.

Melt Flow (熔体流动)—Rate of extrusion of molten resin through a die of specified length and diameter. The conditions of the test (e.g. temperature and load) should be given. Frequently, however, the manufacturer's data lists only the value, not the condition as well.

Melt Index (熔融指数)—The amount of a thermoplastic resin, measured in grams, which can be forced through a specified orifice within ten minutes when subjected to a specified force. (ASTM D-1238)

Modified Resins (改性树脂)—Synthetic resins modified by the incorporation of natural resins, elastomers or other additives, which alter the processing characteristics or physical properties of the basic resins.

Modulus (模量)—Derived from the Latin world meaning "small measure", modulus is the ratio of stress to strain in the linear region of the s-e curve.

Mold (*n*.) (模具)—A hollow form or matrix into which a plastic material is placed and which imparts to the material its final shape as a finished article.

Mold (*v*.) (模塑)—To impart shape to a plastic mass by means of a confining cavity or matrix.

Mold Release (脱模剂)—In injection molding, a lubricant used to coat the surface of the mold to enhance ejection of the molded article or prevent it from sticking to the tool.

Mold Release Agent (脱模剂)—A lubricant used to coat a mold cavity to prevent adhesion of the molded piece when removed.

Molecular Weight (分子量)—The sum of the atomic weights of all atoms in a molecule.

Molecule (分子)—The smallest unit quantity of matter which can exist by itself and retain all of the properties of the original substance.

Monomer (单体)—A relatively simple compound, usually containing carbon and of low molecular weight, which can react to form a polymer by combination with itself or with other similar molecules or compounds.

Nozzle (喷嘴)—In injection molding, the orifice-containing plug at the end of the injection cylinder which contacts the mold sprue bushing and conducts the molten material into the mold.

Opaque (不透明的)—Does not able to transmit light.

Oxygen Index (氧指数)—A flammability test based on the principle that a certain volumetric concentration of oxygen is necessary to maintain combustion of a specimen after it has been ignited.

Pellets (粒料)—Tablets or granules of uniform size, consisting of resins or mixtures of resins with compounding additives which have been prepared for molding operations by extrusion and chopping into short segments.

Photodegradation (光降解)—Degradation of plastics due to the action of light. Pigments General term for all colorants, organic and inorganic, natural and synthetic, which are insoluble in the medium in which they are used.

Plastic (塑料)—A material that contains as an essential ingredient one or more organic polymeric substances of large molecular weight, is solid in its finished state, and, at some stage in its manufacture or processing into finished articles, can be shaped by flow.

Plastic Deformation (塑性变形)—A change in dimensions of an object under load that is not recovered when the load is removed.

Plasticity (塑性)—The ability of a material to withstand continuous and permanent deformation by stresses exceeding the yield value of the material without rupture.

Plasticize (塑炼，塑化，增塑)—To render a material softer, more flexible and/or more moldable by the addition of a plasticizer.

Plasticizer (增塑剂)—A substance or material incorporated in a material (usually a plastic or an elastomer) to increase its flexibility, workability or extensibility.They are usually low-melting solids or high-boiling organic liquids which, when added to hard plastics, impart flexibility. They have varying degrees of softening action and solvating ability resulting from a reduction of intermolecular forces in the polymer.

Plastisol (增塑溶胶)—Mixtures of plasticizers and resins which can be converted to continuous films by applying heat.

Plate-Out (积垢)—An objectionable coating gradually formed on metal surfaces of molds during processing of plastics due to extraction and deposition of some ingredient such as pigment, lubricant, stabilizer or plasticizer.

Polycarbonate Resin (聚碳酸酯)—(PC) A family of special types of polyesters in which groups of dihydric phenols are linked through carbonate linkages.

Polymer (Synthetic) (聚合物，高分子)—The product of a polymerization reaction. The product of polymerization of one monomer is called a homopolymer, monopolymer or simply a polymer. when two monomers are polymerized simultaneously the product is called a copolymer. The term terpolymer is sometimes used for polymerization products of three monomers.

Polymer Structure (聚合物结构)—A general term referring to the relative positions, arrangement in space, and freedom of motion of atoms in a polymer molecule.

Polymerization (聚合反应)—A chemical reaction in which the molecules of a simple substance (monomer) are linked together to form large molecules whose molecular weight is a multiple of that of the monomer.

Processing Aids (加工助剂)—Some processing aids include thixotropic agents, flatting agents, and blocking and anticaking agents.

Processing Methods (加工方法)—The kind of processing (extruding, molding, casting, etc.) techniques recommended by the manufacturer.

Processing Temperature (加工温度)—An average value is given rather than the temperature range often specified by the manufacturer.

Reaction Injection Molding (RIM) (反应注塑)—A process that involves the high pressure impingement mixing of two or more reactive liquid components and injecting into a closed mold at low pressure.

Refractive Index, Sodium D (折射率，钠 D 光)—The ratio of the velocity and light in a vacuum to its velocity in the material.

Reinforced Plastic (增强塑料)—A plastic composition in which fibrous reinforcements are imbedded, with strength properties greatly superior to those of the base resin.

Reinforcement (增强)—A strong, inert fibrous material incorporated in a plastic mass to improve its physical properties.

Resin (Synthetic) (合成树脂)—The term is use to designate any polymer that is a basic material for plastics.

Resin (树脂)—A pseudosolid or solid organic material often of high molecular weight. It has a tendency to flow when subjected to stress, usually has a softening or melting range, and usually fractured conchoidally.

Screw (螺杆)—In extrusion, the shaft provided with helical grooves which conveys the material from the hopper outlet through the barrel and forces it out through the die.

Shear Strength (剪切强度)—The maximum load required to shear the specimen in such a manner that the moving portion has completely cleared the stationary portion. Sheet Sheets are distinguished from films in the plastics industry only according to their thickness. In general, sheets have thicknesses greater than .040".

Sheet (片材)—Sheets are made of continuous phase plastic in a form in which the thickness is very small in proportion to length and width. The thickness is greater than 0.25 millimeters.

Shot (注射)—One complete cycle of a molding machine.

Shot Capacity (注射量)—The maximum weight of material that can be delivered to an injection mold by one stroke of the ram.

Slip Agent (防黏剂，隔离剂)—An additive that provides surface lubrication during and immediately following processing of the plastic material. It acts as an internal lubricant which will eventually migrate to the surfaces.

Solvents (溶剂)—Substances with the ability to dissolve other substances.

Specific Gravity (密度)—The ratio of the density of a material as compared to the density of water at standard atmospheric pressure (1 atm) and room temperature (73℉).

Specific Volume (比容)—The volume of a unit of weight of a material; the reciprocal of density.

Sprue (主流道)—In an injection mold, the main feed channel that connects the mold filling orifice with the runners leading to each cavity gate.

Sprue Gate (主流道)—The passage through which molten resin flows from the nozzle to the mold cavity.

Stabilizer (稳定剂)—An agent used in compounding some plastics to assist in maintaining the physical and chemical properties of the compounded materials at suitable values throughout the processing and service life of the material and/or the parts made therefrom.

Stabilizers & Surface Modifiers (稳定剂和表面改性剂)—Some additives included in this category include antioxidants and antizonants, antistats, biocides and fungicides, heat stabilizers, light, and UV stabilizers and absorbers.

Stiffness (刚性)—The capacity of a material to resist elastic displacement under stress.

Strain (应变)—In tensile testing, the ratio of the elongation to the gage length of the test specimen, that is, the change in length per unit of original length.

Stress (应力)—The force producing or tending to produce deformation in a body measured by the force applied per unit area.

Stress Crack (应力开裂)—External or internal cracks in a plastic caused by tensile stresses less than that of its short-time mechanical strength. Note: The development of such cracks is frequently accelerated by the environment to which the plastic is exposed.

Stress Relaxation (应力松弛)—The decay of stress at a constant strain.

Stress-Strain Curve (应力-应变曲线)—The curve plotting the applied stress on a test specimen versus the corresponding strain. Stress can be applied through shear, compression, flexure, or tension.

Surfactants (表面活性剂)—The use of these chemicals allows the formation of an emulsion or intimate mixture of otherwise incompatible substances by modifying the surface properties and influencing the wetting and flowing properties of liquids.

Tackifiers (增黏剂)—Additives used to enhance the adhesiveness or bonding ability of a material.

Tensile Modulus (拉伸模量)—(Also called modulus of elasticity). The ratio of nominal stress to the corresponding strain below the proportional limit of a material.

Tensile Strength (拉伸强度)—The maximum tensile stress sustained by the specimen during a tension test

Tensile Strength, Break (拉伸断裂强度)—The maximum stress that a material can withstand without breaking when subjected to a stretching load.

Tensile Strength, Yield (拉伸屈服强度)—The maximum stress that a material can withstand without yielding when subjected to a stretching load.

Thermal Conductivity (热导率)—The rate of heat flow under steady state conditions through unit area per unit temperature gradient in a direction perpendicular to an isothermal surface.

Thermoelasticity (热弹性)—Rubber-like elasticity exhibited by a rigid plastic resulting from an increase in temperature.

Thermoforming (热成型)—The process of forming a thermoplastic sheet into a three-dimensional shape by clamping the sheet in a frame, heating it to tender it soft and flowable. Then applying differential pressure to make the sheet conform to the shape of a mold or die positioned below the frame.

Thermoplastic Elastomers (热塑性弹性体)—The family of polymers that resemble elastomers in that they can be repeatedly stretched without distortion of the unstressed part shape, but are true thermoplastics and thus do not require curing.

Thermoplastics (热塑性)—Materials that become soft when heated and solid when cooled to room temperature. This softening and setting may be repeated many times.

Thermoplastics (热塑性塑料)—Resins capable of undergoing a chemical reaction leading to a relative infusible and insolvable state.

Thermosets (热固性)—Materials that may not be reheated and softened again. Once the structural framework is set, these plastics cannot be reformed.

Thermosets (热固性塑料)—Resins or plastic compounds, which in their final state are infusible and insoluble. After being fully cured, thermosets cannot be resoftened by heat.

Transfer Molding (传递成型)—A process of forming articles by fusing a plastic material in a chamber then forcing the whole mass into a hot mold to solidify.

Transition Temperature (转变温度)—The temperature at which a polymer changes from (or to) a viscous or rubbery condition (or from) a hard and relatively brittle one.

Ultimate Elongation (极限伸长)—In a tensile test the elongation at rupture.

Ultimate Strength (极限强度)—Term used to describe the maximum unit stress a material will withstand when subjected to an applied load in a compression, tension, flexural, or shear test.

Vacuum Forming (真空成型)—A method of forming plastic sheets or films into three-dimensional shapes, in which the plastic sheet is clamped in a frame suspended above a mold, heated until it becomes softened, drawn down into contact with the mold by means of a vacuum, and cooled while in contact with the mold. Often used interchangeably with thermoforming.

Vacuum Forming (真空成型)—A process whereby a heated plastic sheet is drawn against a mold surface by evacuating the air between it and the mold.

Viscoelasticity (黏弹性)—This property, possessed by all plastics to some degree, dictates that while plastics have solid-like characteristics such as elasticity, strength and form-stability, they also have liquid-like characteristics such as flow depending on time, temperature, rate and amount of loading.

Viscosity (黏度)—A measure of the resistance to flow due to internal friction when one layer of fluid is caused to move in relationship to another layer.

Weathering (天候老化)—A broad term encompassing exposure of plastics to solar or ultraviolet light, temperature, oxygen, humidity, rain, snow' wind, and air-borne biological and chemical agents.

Yield Point (屈服点)—In tensile testing, yield point is the first point on the stress-strain curve at which an increase in strain occurs without an increase in stress.

Yield Strength (屈服强度)—The stress at which a material exhibits a specified limiting deviation from the proportionality of stress to strain.

Young's Modulus (杨氏模量)—The ratio of tensile stress to tensile strain below the proportional limit.

Vocabulary

A

abbreviation	缩写	C6
abrasion	磨损	A4R
abrasive	研磨剂，研磨的	F3
abrasive wear	磨损	D5
absorbent	吸收剂	A4R
accelerator	促进剂	D1
accompany	伴随	D5
accumulation	积聚，堆积	E7R
acetate	醋酸酯、乙酸酯、醋酸盐	A1
acetic	醋酸的，乙酸的	A3
acetylene	乙炔，电石气	C3
acrylic	丙烯酸的	B3R
acrylic rubber (ACM)	丙烯酸酯橡胶	D2
acrylonitrile	丙烯腈	B1
activate	活化	D3
activated	有活性的	A2
addition	加成	A2
additive	添加剂	A4
adhesive	胶黏剂	A1R
adipate	己二酸	B2R
adipic	脂肪的	B2R
adjacent	邻近的，接近的	D3
adjustable	可调整的	E1
aerobic	有氧气的	D3
aesthetic	美学的，审美的	B2R
agent	介质，剂	C4
aggregate	聚集，集合	D5
aggregation	聚集物，聚集（态）	C2R
aggressive	活性高的	E6
aging	老化	B1
agitation	搅拌	A3R
albumin	白蛋白	A1R
alcohol	醇，乙醇	A1
aldehyde	醛，乙醛	C5
aliphatic	脂肪族的，脂肪质的	C3
aliphatic	脂肪族的	C6
alkali	碱，碱性的	A4R
alkaline	碱性的，碱的	C5
alkane	烷烃	C1R
alkylation	烷基化	E6
alkyllithium	烷基锂	E2
allbut	几乎，差一点	E4
allergic	毒性	E7R
alloying	合金	D2R
allylic	烯丙基的	E6
alternative	二中选一	B1R
alumina	氧化铝(亦称矾土)	C1R
aluminum	铝	A4R
amber	琥珀	A1R
ambient	环境温度，室温	C3R
amide	酰胺	A3
amine	胺	A3
amorphous	无定形的，非晶的	A4
amortize	分期清偿	F2
anhydride	酐	C6
anionic	阴离子的	A3
antiager	防老剂	D4
antidegradant	防老剂，抗降解剂	D1
antioxidant	抗氧剂	B1
antiozonant	抗臭氧剂	D1
antistatic	抗静电的	D5
apolar	非极性的	D5R
appeal	要求	B2R
appearance	外貌，外观	E1
appliance	用具，器具	E4

application	应用，申请，运用	A4	block	嵌段	D2R
approach	方法，步骤，途径	E7R	bloom	喷霜	E7R
approximately	近似地，大约	D5	boiling	沸腾的	F2R
aqueous	水的，水溶液的	A3R	bond strength	黏结强度	E6R
arbitrary	任意的，武断的	D5	bonding	粘接，接合	E6R
architecture	构造	A2R	boundary	分界线	A1
aromatic	芳香的	C6	branch	分支	A1
array	大批	E4R	breakdown	破坏	E5
asbestos	石棉	A4R	bristle	短粗纤维	F1R
asphalt	沥青	A1R	brittle	易碎的，脆弱的	B2R
assortment	分类	A3R	broaden	放宽，变宽，加宽	E3R
attribute	属性，品质，特征	D5	bromide	溴化物	C2
autoacceleration	自加速	A3R	brominate	溴化	E6
autoclave	高压容器，高压釜	C1	bromine	溴	C4
autoclave	硫化罐	E3	bromobutyl	溴化丁基胶	E6
azelate	壬二酸酯	B2R	buffing	磨削，抛光	F3

B

			bulk	大批，本体	A3R
backbone	主链	E2	bulky	大的，容量大的	C2
balata	巴拉塔胶	E2	butadiene	丁二烯	D2
band	包辊	E3R	butyl	丁基	D2
barium	钡	D5			

C

base	碱	A3	ca. (=about)	大约	E5
batch	间歇	E4R	cadmium	镉	B2R
batch	一批	F3R	calcine	烧成石灰，煅烧	B3R
belie	掩饰	F1R	calender	压延,压延机	B3
benzaldehyde	苯甲醛	C1	calenderability	压延性能	E5
benzene	苯	C1	carbohydrate	碳水化合物，糖类	A1R
benzophenone	苯甲酮	B1	carbon	碳	A1
benzotriazole	苯并三唑	B1	carbonate	碳酸盐	D5
benzoyl	苯(甲)酰	C1	carboxylate	羧酸盐，羧酸酯	E4R
biaxial	双轴（的）	C2	carboxylic	羧基的	A3
biocide	生物杀灭剂	C4R	carcinogenic	致癌物(质)的	A3R
biodegradable	生物所能分解的	C6R	cast iron	铸铁	A4R
bisphenol	双酚	D4	casting	浇铸	D2
bitumen	沥青	E5	catalysis	催化作用	D4
blank	胶坯	F3	catastrophic	悲惨的，灾难的	C4R
blend	混合，共混	A3	category	种类	D2
blend	共混，并用，并用胶	D1	cation	阳离子	A3

cavity	模腔	F2		commercial	商业的，贸易的	E1R
cellulose	纤维素	A1R		commercialize	使商业化，使商品化	B1R
cellulosics	纤维素	C6R		communication	信息，交通，通信	D1R
centrifugal	离心的	F1		comonomer	共聚单体	E5
ceramic	陶瓷；陶器的	A4R		comparatively	比较地，相当地	E5R
chain reaction	链式反应，连锁反应	D4		comparison	比较，对照	E2R
chain scission	断链	C4R		compatibility	相容性	E4
characteristic	特性，特征	D5		complicated	复杂的，难解的	D3R
charcoal	活性炭	C1R		component	成分	D2R
chemical	化学的；化学药品	A4R		composition	成分	B1
chemical attack	化学腐蚀	D4R		compounder	配料员，配方人员	E4
chloride	氯化物	A1		compression	模压	F3
chlorinate	氯化	E6		comprise	包含，由…组成	A2
chlorine	氯	A1		concentration	浓度，浓缩	A3R
chlorobenzene	氯苯	C1		condensate	缩合物	D4
chlorobutyl	氯化丁基胶	E6		condensation	缩合	A3
chloroprene	氯丁二烯	D2		conductive	传导的	E1
chop	切，砍	A4		conductivity	电导率	A4R
chromic	铬的	C1R		conductor	导体	E3
cis-	顺式	D2		confer	授予，赠与	D5
civilization	文明，文化	D1R		configuration	构造，结构，构型	A4R
clamp	夹子，夹具，夹钳	F3R		conjugated	共轭的，成对的	F1
classify	分类，分等	A2		consensus	一致同意	C6R
clay	黏土，陶土	D1		consequence	结果	D5R
clean out	清除，打扫干净	B3		consequent	随之发生的	D5
coagent	活性助剂	D3		considerably	相当地	E3
coagulant	凝聚剂	E4R		constant	常数，恒量	A3R
coagulation	凝聚	E3		constituent	成分	F3R
coalescence	合并，凝结	A3R		constraint	约束，强制	A4R
coat	涂上，包上	F2R		construction material	结构材料	C5R
coaxial	同轴的，共轴的	C1		consumption	消费，消费量	D2
cobalt	钴(Co)	C1R		contamination	过敏的	E7R
coefficient	系数	A4R		conventional	常规的，传统的	D3
coffin	棺材	D1R		conversion	转化	A2
collectively	全体地，共同地	D4R		convert	使转变，转换	D1
colorfast	不退色的	A4R		copolymer	共聚物	A3
comb polymer	梳型高聚物	A2R		copolymerization	共聚合	A3
come with	伴随…发生	D5R		corrosion	侵蚀，腐蚀状态	C3

corrugated	波纹的	C1R		delayed-action	迟效性	D3
coupling agent	偶联剂	D5R		demanding	过分要求的, 苛求的	E7
covalent	共价的	A1		denature	使变性	A1R
cracking	龟裂, 裂口, 裂纹	D1		dendrimer	超支化高分子	A2R
cradle	摇篮	D1R		dendronized polymer	树形高分子	A2R
creep	蠕变	D3		denier	旦尼尔	A4
cresol	甲酚	C5		denotes	指示, 表示	D5
criteria	标准	C6R		dependent	依赖的, 由…决定的	D1R
critical	关键的	E6		depolymerize	(使)解聚	A1R
crosslink	交联	A2		deposit	堆积物, 沉淀物	E7R
crosslinking	交联	C1R		derivative	衍生物	C4
cryogenics	低温	F3R		designate	指定	D5
crystalline	结晶的	A4		deterioration	变坏, 退化	D1
crystalline state	晶态	E2R		determine	决定, 确定	E5R
crystallinity	结晶度	A2R		devulcanization	脱硫(作用)	D2R
crystallite	微晶	E6R		dialkyl	二烷基	D3R
cure	固化	C6		diameter	直径	F3R
cushioning	减震, 缓冲	D1R		diamine	二胺	C6
cutting	切割, 冲裁	F3		dianiline	二苯胺	C6
cyclic	环状的	D3		diaryl	二芳基	D3R
cyclohexane	环己烷	C1R		diazomethane	重氮甲烷	C1
cylinder	机筒	F2		dicumyl peroxide	过氧化二异丙苯(DCP)	C1R
				die	口模, 口型	F1

D

damage	损害, 伤害	D4R		dielectric	电介质, 绝缘体	A4R
dampen	阻尼	E6		dielectric constant	介电常数	A4R
damping	阻尼, 衰减	D2		diene	二烯(烃)	D2
decompose	分解	A2		diffusivity	扩散能力	F2
decyl	癸基	B2R		digit	阿拉伯数字	D5
definition	定义, 解说	D2R		diglycidyl	缩水甘油基	C6
deflashing	修边, 除边	F3		diisooctyl	二异辛基	E5R
deformation	变形	D2		dilauryl	二月桂基	C2
degradability	降解性	C6R		diluent	稀释的; 稀释剂	B2
degradation	降解	B1		dilute	淡的, 稀释的	E3
degrade	(使)降级	A1R		dimer	二聚物	A3
degree of branching	支化度	A2R		diolefin	二烯	E6
degree of crosslinking	交联度	A2R		dioxide	二氧化物	B1
dehydrochlorination	脱氯化氢	C3		dip	浸渍	D2R
delay	耽搁, 延迟	E5R		dip	浸, 蘸, 沾	F2R

diphenylamine	二苯胺	D4	elastomeric	弹性体的	D2
disadvantage	不利，缺点	D5R	element	元素，成分	A1
discoloration	变色，污点	B2R	elevated temperature	高温	D2R
discrete	不连续的，离散的	B3	elimination	除去，消除	A2
dispersion	分散	E5R	elongation	伸长	B2
displace	移置，取代，置换	E6	emanate	发出，发源	A2R
dissipate	驱散，(使)消散	A3R	emerge from	自…出现	F3R
dissipation	消散，分散	F1	emphasize	强调，着重	C5R
distinct	明显的，独特的	D4R	emulsifier	乳化剂	E3
distinction	区别，差别	B2	emulsion	乳液	A3R
distortion	扭曲，变形	F2	encapsulation	包胶，封装	C6
distribution	分布	E3R	engender	造成	C2R
disulphidic	二硫的	D3	entanglement	缠结	A2R
dithiocarboxylic	二硫代碳酸的	C2	environmental	环境的，周围的	E1R
dithiopropionate	二硫代丙酸酯（盐）	C2	epichlorohydrin	环氧氯丙烷	C6
dithocarbamate	二硫代氨基甲酸盐	D3	epoxy	环氧树脂	C6
dodecenylsuccinic	十二碳烯基琥珀酸的	C6	equivalent	相等的，相当的	D3R
dodecyl	十二(烷)基	E5	ester	酯	A3
domain	范围，领域	A1	ethanol	乙醇	C1R
dominate	支配，占优势	B2	ether	醚	C6
drainage	排水装置	C5R	ethyl	乙基	B2
drape	悬垂性	A4	ethylene	乙烯，亚乙基	D2
draw	汲取，领取，吸引	E7R	evaporation	蒸发(作用)	F2R
dryblend	干混料，干混合	C3R	exceed	超越，胜过	E2R
ductility	延展性	C1R	excessive	过多的，额外	B3
durable	持久的，耐用的	E6	exothermic	放热的	C1
duration	持续时间，为期	A2	exploit	使用	E7R
durometer	硬度计	D5R	extend	填充	E3
dye	染料，染色	A1R	extender	增容剂	B2
dynamic	动态的	E6R	extensibility	延长性，伸长	A4

E

			extension	延长，伸长	D2
ebonite	硬橡胶	A4	external	外部的，外面	D4R
economy	经济，节约	E2	extinguishing	熄灭	B3R
efficient	有效率的	D3	extract	抽提	E7R
elastic	弹性的	D1	extremely	极端的，非常的	C3
elastic band	橡皮筋	D5R	extrude	挤出	A4R
elasticity	弹性	A4	extruder	压出机	E4
elastomer	弹性体	A1R	extrusion	挤出	B3

F

fabric	构造	D1R
fabricate	制作, 构成	C4
fabrication	制作, 装配	D3
facility	设备, 工具	C1
fall into	分成, 属于	F3
fatigue	疲乏, 疲劳	A4R
fatty acid	脂肪酸	D3
feed	进料, 喂料, 流入	F3
fibrous	纤维状的	D5
filament	细丝	A4
filler	填料, 填充剂	A4
film	薄膜	A2R
filter	过滤	C1R
finish	光洁度	A4R
finishing	修饰, 整理	F3
fire retardance	阻燃性	A4R
flame-retardant	阻燃剂	C4
flammability	易燃, 可燃性	E5
flax	亚麻, 麻布, 亚麻织品	A1R
flex	曲挠, 多次弯曲	D1
flexible	柔性的, 柔软的	B2
fluctuation	波动, 起伏	C4R
fluffy	松散的, 蓬松的	C3
fluidized-bed molding	流化床模塑	F2R
fluorescent	荧光的	C4
fluorine	氟	A1
fluorocarbon	碳氟化合物	B1
foam	泡沫; 发泡	B2
forcibly	强制地, 有力地	A4
foremost	首要地, 首先	E7R
formaldehyde	甲醛, 蚁醛	C5
formation	形成, 构成	E3R
formula	化学式, 公式	A3
formulating	配制, 配料	C4
formulation	配方, 配料, 配合,	D1
foster	鼓励	C6R
fraction	级分, 组分	E3R
freeze	(使)结冰, (使)冷冻	E6R
frictional	摩擦的, 摩擦力的	B3
fuel	燃料	E2
functional group	官能团	A1R
functionality	官能度	A2R
fungicide	杀菌剂	B3R
furfural	糠醛	C5
furnish	供应, 提供, 装备; 供给	D1R
furniture	家具	E1R
fusion	熔化, 熔解	B3

G

gap	缺口, 间隙	F1
gas permeability	透气性	D2
gasket	垫圈, 衬垫	D5
gauche	左右式的, 旁式的	B2
gel	凝胶	A2R
gelatin	明胶, 凝胶	A3R
glazing	玻璃装配业, 玻璃窗	F1R
glycol	乙二醇	B2R
goods	制品	D5
granule	小粒, 颗粒	C2R
grave	墓穴, 坟墓	D1R
green strength	生胶强度	E3R
green tire	生胎, 胎胚	E3R
grind (ground, ground)	磨(碎)	C1
grip	紧握, 紧夹, 抓住	D5
grommet	垫圈	E4
guanidine	胍	D3
gum	纯胶	E3
gunk	黏性物质	C5
gutta percha	古塔波胶	E2R

H

half life	半衰期	D3R
halide	卤化物, 卤化物的	C4
halogen	卤素	E6
halogenate	卤化	E6
halt	停止, 中断	B3
hand	手感	A4

handle	运用，处理，操作	E7R	inactive	无活性的	D3
haul	拖拉，拖运	D1R	incompatible	不相容的	E5R
hearse	柩车，灵车	D1R	incorporated	合成一体的	F3R
hefty	有力的，很重的	F2	indentation	凹痕	D5R
helix	螺旋，螺旋状物	B2	indicator	指标	A1
heterocyclic	杂环的	B1R	indispensable	不可缺少的，绝对必要的	D1R
heterogeneous	非均相的	A3R	inertness	不活泼，惰性	A4R
Hevea brasiliensis	巴西三叶胶树	D2	inferior	差的，次的	E6R
hexahydrophthalic	六氢化邻苯二甲酸的	C6	infinite	无限的，无数的	D5R
hexamethylene	己基	C6	inherently	天生地，固有地	D5
hexamethylenetetramine	六亚甲基四胺	C5	inhibit	抑制	B1
hexane	己烷	C2	inhibitor	抑制剂	D4
hexyl	己基	B2	initiate	引发	D3R
hinder	阻碍，打扰	D4	initiation	引发	A3
homogeneous	均一的，均匀的	E1	initiator	引发剂	A3
homogenize	均质化，使均质	C2R	injection	注射	E3
homopolymer	均聚物	A3	inlay	镶嵌	E6R
hopper	料斗	F2	innerliner	气密层	E6
hydraulic	液压的，水压的	E4	innertube	内胎	E6
hydraulically	液压	F2	insert	插入，嵌入	D3
hydride	氢化物	C1R	insulating	绝缘的	E1
hydrocarbon	烃，碳氢化合物	B1	integral	构成整体所必需的	C6R
hydrochloric	氯化氢的，盐酸的	C3	integrate	使成整体，结合	C6R
hydrogen	氢	A1	integrity	完整性	B1R
hydrolyse	水解	A1	interact	互相作用	B2
hydroperoxide	氢过氧化物，过氧化氢物	D4	interconnect	使互相连接	A1
hydroxyl	羟基	A2	interfere	干涉，干预，妨碍	D4
hysteresis	滞后作用	E4	intermediate	中间的	D3
			intermesh	使互相啮合	F1

I

illustration	图表，插图	D4R	intermittent	间歇的，断断续续的	E1R
impact	冲击	B1	intermolecular	分子间的	A1R
impart	给予，传授	E2	intramolecular	分子内的	A1R
implant	植入	A4R	intricate	复杂的，错综的	A4R
impregnate	浸渍	C5	ion	离子	D4
in comparison with (to)	与…比较	E2R	ionic	离子的	A3
in contrast to	和…形成对比	E5	irregular	不规则的，无规律的	F2R
in the order of	大约	D5	irregularity	不规则，无规律	E5

isobutene	异丁烯	C1R		manufacturer	制造业者, 厂商	D1
isobutylene	异丁烯	D2		mass	质量	A1
isocyanate	异氰酸酯	A2		masterbatch	母炼胶	E3
isomeric	异构的	E4R		masticate	塑炼, 破料	C3
isoprene	异戊二烯	D2		mat	垫子	C2
isotactic	等规的, 全同立构的	C2		material	材料, 原料, 物资	D1R
issue	论点, 问题	D5R		matrix	基质, 基体	C2R

J K L

				meat grinder	绞肉机	F1R
jar	震动, 刺耳声, 争吵	D1R		mechanism	机理, 历程	A2
jolt	摇晃	D1R		mechanistic	机械论的	D4R
kaolin	高岭土, 瓷土	A3R		medical	内科的, 医学的	D3R
ketone	酮	D4		medium	介质	A2
kinetics	动力学	A2		melamine	三聚氰胺	B2
knockout pin	顶销, 顶杆	F2		melt	(使)熔化	A2
labile	不安定的	D3R		melt viscosity	熔体黏度	E2R
ladder	梯形高分子	A2R		melting point	熔点	E2R
lamellar	片状的, 层状的	D5		membrane	膜, 隔膜	E7
laminate	层压板	C5		mercaptan	硫醇	E5
lanthanum	镧(La)	E2		mercaptide	硫醇盐	B2R
laureate	佩戴桂冠的	A1R		mesh	啮合	F3R
layer	层	D4		metal oxide	金属氧化物	D3R
lead	铅	B1R		methacrylate	甲基丙烯酸酯（盐）	B1
limestone	石灰石	C1		methanol	甲醇	C2
linear	线型的	A1		methyl	甲基	B1
literally	逐字地	D1R		methylene	亚甲基	C6
loading	加入量	E3		mica	云母	A4R
long term	长期	E5R		micrometer	微米	A3R
looter	掠夺, 抢劫, 强夺	D1R		microporousness	微孔性	C2R
lower	降低, 减弱	B3		microwave	微波	E3
lubricant	滑润剂	B3		migrate	移动, 移往	E7R
lubricity	润滑	B3		milk jug	奶壶	A2R

M

				mill	炼塑	B3
macromolecule	大分子, 高分子	A1R		mineral	矿质的, 无机的	D5
macroscopic	宏观的	B2		miniature	缩影, 微型的, 缩小的	A3R
magnesium	镁(Mg)	D5		miscellaneous	混杂的	E3
magnitude	大小, 数量, 量级	A4R		mode	方式, 模式	E5R
malleable	有延展性的	A4R		modification	改性	D3
manganese	锰(Mn)	D4		modifier	改性剂	E2

modifing resin	改性树脂	C4R
modulus	模量	A4
modulus	定伸强度	E5R
molar	摩尔的	A1
mold	模型；模塑，模压	A4R
molding	模塑，模压，成型	F2
molecule	分子	A1
molybdenum	钼	C1R
monofilament	单(根长)丝，单纤(维)丝	F1R
monomer	单体	A1
monosulphidic	单硫的	D3
montmorillonite	蒙脱石	C2R
mop	(抛光用)毛布轮	F3
moulding	成型	F3
mounting	减震器	D5
mulch	覆盖，覆盖物	C1
multidomain	多畴	C2R
myriad	无数的，种种的	D1R

N

nanometer	纳米	C2R
necessitate	成为必要	D1
negative	负，阴性的	C3
network	网状物	A1
neutralize	压制	A3
nickel	镍	C1R
niobium	铌(Nb)	E2
nip	辊距	F3R
nitrate	硝化	C4
nitrile rubber(NBR)	丁腈胶	D2
nitrogen	氮	A1
nomenclature	命名法	C1
nonyl	壬基	E7R
not less than	至少	D3
notably	显著地，特别地	E7R
novolac	线型酚醛树脂	C5
nozzle	喷嘴	F2
nylon	尼龙	F2R

O

obsolete	陈旧的，荒废的	C5R
occasion	场合，时机，机会	E6R
octyl	辛基	B2
odor	气味	C4
oligomer	低聚物	B2
opacity	不透明性	B3R
optical	光学的	C4
optimize	使最优化	C6R
optimum	最适宜的	D3
organic	有机的	A1
organo	有机金属的	B2R
organosol	稀释增塑糊，有机溶胶	C3R
orient	取向，定向	E6R
origin	起源，由来，起因	E7R
outlay	费用	F2R
outstanding	突出的，显著的	C5R
oven	烤箱，烤炉，灶	F2R
oxidation	氧化	B1
oxidative	氧化的，具有氧化特性的	C2
oxygen	氧	A1
ozone	臭氧	D1

P

packer	垫片	E4
paint	油漆，颜料，涂料	C5R
paraffin	石蜡	C1
parallel	相应，平行	B2R
parameter	参数，参量	B2
parison sag	型坯下垂	F2R
particulate	微粒；微粒的	D1
passenger car	乘用车	D5R
pastry	面粉糕饼，馅饼皮	F3R
pellet	粒料	F1
pendent	悬挂的，侧基的	E2
penetrate	渗透，浸染	F1
peptization	塑解	E5
performance	性能	D1

perish	腐烂, 枯萎, 毁坏	E1R		polymer	聚合物	A1
permanent	永久的, 持久的	A4R		polymerization	聚合	A1
peroxide	过氧化物	D3R		polymerize	(使)聚合	E5
peroxy	过氧	C1		polymethylene	聚亚甲基	C1
peroxy radical	过氧自由基	D3R		polyolefin	聚烯烃	B3R
peroxyester	过氧化酯	D3R		polyphenol	多元酚	D4
peroxyketal	过氧化酮	D3R		polypropylene	聚丙烯	B1
perpendicular	垂直的, 正交的	F2R		polysaccharide	多糖, 聚糖, 多聚糖	A1R
petroleum	石油	C3		polystyrene	聚苯乙烯	B1
pharmaceutical	药物, 制药(学)上的	E6		polysulphide	多硫化物	D2
phenol	苯酚	D4		polysulphidic	多硫的	D3
phenolic	酚的, 石炭酸的	C2		polyurethane	聚氨酯	A2
phenyl	苯基	D4		polyvinyl alcohol	聚乙烯醇	C6R
phosphate	磷酸盐(酯)	C2		popcorn-like	米花状	C3R
phosphorus	磷	A1		porcelain	瓷器, 瓷	A4R
phthalate	邻苯二甲酸酯 (盐)	B2		portion	部分	E5R
phthalic	邻苯二甲酸的	B2		position	安置	E6R
pigment	颜料	A4		post-cure	二次硫化	E6R
plasticization	增塑, 塑化, 塑炼	B2		potential	潜在的, 可能的	D4R
plasticize	增塑	D1		pourable	可倾倒的	E4R
plasticized	增塑的	F1R		p-phenylenediamine	对苯二胺	D4
plasticizer	增塑剂	A4		pragmatic	实际的, 注重实效的	C6R
plastisol	增塑溶胶, 增塑糊	C3R		precipitate	使沉淀	D5
plumbing	管	E4		precrosslinked	预交联的	E5
plunger	柱塞	F2		predetermine	预定, 预先确定	B3R
pneumatic	充气的	D2		predominant	主要的	D3
pocket	袖珍的, 小型的	D5R		preferentially	择优地	B2
polar	极性的	D5R		preforming	预成型	F3
polarity	极性	E4		premature	未成熟的	C2R
polish	光泽, 上光剂	C1		prepolymer	预聚物	C6
polyamine	多元胺	C6		preservation	保存	C5R
polybasic	多元的	C6		press	平板硫化机	F3
polybutadiene	聚丁二烯	E2		previously	先前, 以前	D1R
polycondensation	缩聚反应	B2		primarily	首先, 起初, 主要地	D4
polyester	聚酯	B1		principal	主要的, 首要的	D3R
polyether	聚醚, 多醚	B2		principle	原理, 法则, 原则	A1R
polyisoprene	聚异戊二烯	D2		proccessability	加工性	B2
polylactic acid	聚乳酸	C6R		process	过程, 加工, 操作	D1

English	Chinese	Ref
processible	可加工的	E5
product	产品, 产物, 乘积	A1
profile	剖面, 轮廓	F1
profiled part	异型配件	E5
profound	深刻的, 深奥的	C3
prone to	倾向于…	E5R
pro-oxidant	助氧化剂	E6R
propagation	增长	A3
proportion	比例, 部分	D3
propylene	丙烯	C2
prosperity	繁荣	D1
protective	保护的	D4
proteinaceous	蛋白质的, 似蛋白质的	A1R
provided	倘若	E6R
proximity	接近	E5R
pulverize	研磨成粉	C5
punching	冲孔, 冲眼	F3

Q R

English	Chinese	Ref
quality	质量, 性质	E3
quality assurance	质量保证	F3R
quantitative	定量的	A4
radiation	辐射, 放射	B1
radiator	散热器, 水箱	E7
radical	根本的, 激进的	E7R
ratio	比, 比率	A4
raw	未加工的, 生疏的	D4
reactant	反应物	C1R
reciprocating	往复的	F1
reclaimation	再生, 回收	F1
reclaimed rubber (RR)	再生胶	D2R
recovery	恢复	D2
recyclability	再循环性, 再回用性	C6R
recycle	再循环, 反复应用	D2R
redox	氧化还原作用	D4
reflection	反映	D5R
reflux	回流	C3R
regulate	管制, 控制	D3R
reinforce	补强, 加固	A4
reinforcement	加强, 补强	D5
relaxation	应力松弛	E5R
release	剥离	B3
release	释放	D2
remains	残余	D3R
removing	消除, 除去	D4
replenish	补充	E1R
repulsion	排斥, 推斥	C3
resemblance	类同之处	A3R
residual	剩余的, 残留的	D3R
residue	残余物	E7R
resilience	回弹性	D3
resiliency	回弹性	E2
resilient	弹回的, 有回弹力的	E1R
resin	树脂	A1R
resistivity	抵抗力, 电阻	E7
resole	体型酚醛树脂	C5
retain	保持, 保留	D2R
retard	延迟, 阻止	B1
retardant	延缓(作用)剂	D1
reversible	可逆的	E6R
reversion	返原, 回复, 复原	D3
rigid	刚性的	B3
rigidity	坚硬, 刚性, 刚度	A4
ring	链状	A2R
rotational molding	旋转模塑	F2R
rotor	转子	F3R
roughness	粗糙, 粗糙程度	E1
route	路线, 路程	D4R
runner	分流道	F2

S

English	Chinese	Ref
sacrifice	牺牲, 献身	D2R
sap	树液	E1R
scavenger	清道夫	B1R
scheme	方案	A2
scheme	安排, 计划, 方案	E1
scorch	焦烧	E5R

scrap	小片, 废料	D2R		solubility	溶解性, 溶度	A3R
screw	螺杆	F1		soluble	可溶的, 可溶解的	E5R
seal	密封圈	D5		solution	溶液, 溶解	A3R
sealant	密封剂	E6		solvation	溶剂化（作用）	B2
sebacate	癸二酸盐(或酯)	E5R		solvent	溶剂	A3
sebacic	癸二酸的	B2R		span	跨度, 范围	E4R
secondary	次要的, 第二的	E5R		sparingly	部分地	E7R
segment	链段	B2		specific gravity	相对密度，比重(旧称)	A4R
self-reinforce	自补强	E2R		spherical	球的, 球形的	D5
semi-	半	D3		spherulite	球晶	C1
semi-reinforcing	半补强	D5		spin（span spun）	纺, 纺纱	A4
separate	个别的, 单独的	D1		splinter	裂片, 裂成碎片	C4R
serendipitous	偶然发现的	C1		split line	分型线	F3
set	永久变形	D3		sprue	主流道	F2
shear	剪切	B3		spur	鞭策, 刺激, 驱策	C5R
sheet	压片	F3		stability	稳定性	B3
shelf ageing	储存老化	D4R		stabilizer	稳定剂	B1
shrinkable	会收缩的	F1R		stable	稳定的	E2
shrinkage	收缩	E3R		staining	污染, 着色	D4
side chain	侧链, 支链	A2R		standardized	标准的	D5
sidewall	胎侧	E2		standpoint	立场, 观点	A4
signify	表示, 意味	E4		staple	短纤维	A4
silence	使沉默, 使安静	E1		star polymer	星形高聚物	A2R
silica	硅土	C1R		static	静态的, 静电的	B3R
silica	白炭黑	D1		static modulus	静态模量	E6R
silicate	硅酸盐	D5		steam	蒸汽	E3
silicon	硅	A1		steam-strip	汽提	C3R
silicone	聚硅氧烷, 硅酮(旧称)	D2		stearic acid	硬脂酸(SA)	D3
simultaneously	同时地	F1R		stereochemistry	立体化学	A3
slippage	滑动, 滑移	B3		stereoregular	立构规整的	E2
slurry	浆, 泥浆, 料液	C2		stereoregularity	立构规整性	A3
slush molding	搪塑	F2R		steric hidrance	位阻	C3
smelly	发臭的, 有臭味的	E1		sterile	消过毒的	D1R
social benefit	社会利益	C5R		stiffen	刚性, 使变硬	D1
sodium	钠	C1		stiffness	坚硬, 硬度	C2
soften	(使)变柔软	D1		stir	搅动, 搅拌	A3R
sole	鞋大底	D5		stock	胶料	E3
solidify	(使)凝固	A2		stopper	塞子	E6

strain	应变	A4	tatrachoride	四氯化物	C1R
strain-crystallize	应变(诱导)结晶	E2R	technique	技巧，方法	E3
strategically	战略上	F1	tenacity	强力	A4
streak	条纹	B1R	tensile	可伸长的，可拉长的	C4
streamlined	最新型的，改进的	C6R	termination	终止	A3
strength	强度	A2R	terpolymer	三元共聚物	D2
strengthen	加强，巩固	C5R	tertiary	叔，第三的，第三位的	C2
stress	应力	A4	tetramine	四胺	C6
stretching	拉伸，伸长[展]	E2R	thermal	热的，热量的	A2
strip	条，带	F3R	thermal form	热成型	A4R
styrene	苯乙烯	D2	thermodynamic	热力学的	B2
substantial	实质的，真实的	D2R	thermoplast	热塑性塑料，热塑性	B2
substituted	取代的	B1	thermoplastic	热塑性的；热塑性塑料	C3
substrate	底层，基层，被粘物	E4	Thermoplastic elastomer (TPE)	热塑性弹性体	D2R
successive	连续的	A3	thermoplastic rubber (TPR)	热塑性橡胶	D2R
sulfonate	磺化	C4	thermoplastics	热塑性材料，热塑性塑料	A2
sulfur	硫黄	A1			
sulphate	硫酸盐	D5	thermoset	热固性材料，热固性塑料	A2
sulphenamide	次磺酰胺	D3			
sulphide	硫化物	D3	thiazole	噻唑	D3
summarize	概述，总结	E4	thiocarbamate	硫代氨基甲酸盐	E6
superficial	表面的，肤浅的	A3R	thiuram	秋兰姆	D3
superimpose	添加，双重	E6R	threshold	开始，开端	D4
superior	较高的，上级的	E3R	thud	碰击声，重击	C4
surgical	外科的；外科手术	A4R	titanium	钛	B1
surgical	手术上的，外科的	D3R	toughen	(使)变坚韧	E2
susceptible	易受影响的	A3	toughness	韧性	A2R
suspension	悬浮，悬浮液	A3R	toxic	有毒的	A3R
sustained	持续不变的	E6	toxicity	存放，堆积	E7R
swell	溶胀，增大	E3	trace	微量	D3R
synchronous	同时的，同步的	D4R	traction	牵引	E2
synergism	合作，并用	E7R	trans-	反式	B2
synergistic	协同	B1R	translucent	半透明的	C2
synthesis	合成	A2	transparent	透明的	A4R

T

tacticity	立构规整性	C2	transportation	运输，运送	D1R
take advantage of	利用	E2R	tread	胎面	E2
talc	滑石，云母	B3R			

treatment	加工，处理	D2R	version	类型	E6
triamine	三胺	C6	vessel	容器，器皿	C3
triethyl	三乙(烷)基的	C1R	vibration	振动，颤动，摇动，摆动	D1R
triethyl	三乙基的	C6			
tubeless	无内胎的	E6	vigorously	精神旺盛地	C5R
tumble	使滚翻，弄乱	F3	vinegar	醋	A1R
tumult	吵闹，骚动，混乱	D1R	vinyl	乙烯基	A1
turf	草根土，草皮	C2	vinyl acetate	乙酸乙烯酯	C3R
twist	捻，编织	A4	viscosity	黏度，黏性	A3R
typically	代表性地，主要地	E2	viscous	黏性的，胶黏的	F1
			visualize	想象	F1R
U			volatility	挥发性	E6R
ultraviolet	紫外线的	C2	vs. (versus)	对	E6R
unavoidable	不可避免的	D4R	vulcanization	硫化	A2R
undercure	欠硫	E7R	vulcanize	硫化	D1
undergo	经历	A2			
undergoing	经历，遭受，忍受	C4R	**W X Y Z**		
undue	不适当的	D2R	wanting	欠缺的，没有的	D3R
uniform	均匀的，一致的	F1	waterproof	防水的，不透水的	C5R
universally	普遍地，全体地，到处	E4R	wearability	穿着性能，耐磨损性	A4R
unreacted	未反应的	A2	weathering resistance	耐候性	A4R
unsaturated	不饱和的	A2	whiting	碳酸钙，白垩	D5
unsaturation	不饱和	D2	wholly	整个，统统，全部	D1R
unzipping	拉开…上的拉链	C4R	withstand	抵挡，经受住	E4
unzipping effect	拉链式解聚合	C4R	workhorse	主力军	E4
UPVC	未增塑聚氯乙烯	C5R	xanthogen	黄原酸	E5
Usher	引导，展示	D1R	zinc	锌	B1
			zinc oxide	氧化锌（ZnO）	D3
V					
Valve	阀	F2			

注：单词后面的注释为单词所在单元，如标为"B2"为"PART B Unit 2"部分的单词，"B2R"则为"PART B Unit 2"后的"Reading Material"的单词。